Studies in Computational Intelligence

Volume 700

Series editor

Janusz Kacprzyk, Polish Academy of Sciences, Warsaw, Poland
e-mail: kacprzyk@ibspan.waw.pl

Tomasz Traczyk · Włodzimierz Ogryczak
Piotr Pałka · Tomasz Śliwiński
Editors

Digital Preservation:
Putting It to Work

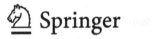 Springer

Editors
Tomasz Traczyk
Institute of Control and Computation
  Engineering
Warsaw University of Technology
Warsaw
Poland

Włodzimierz Ogryczak
Institute of Control and Computation
  Engineering
Warsaw University of Technology
Warsaw
Poland

Piotr Pałka
Institute of Control and Computation
  Engineering
Warsaw University of Technology
Warsaw
Poland

Tomasz Śliwiński
Institute of Control and Computation
  Engineering
Warsaw University of Technology
Warsaw
Poland

ISSN 1860-949X   ISSN 1860-9503   (electronic)
Studies in Computational Intelligence
ISBN 978-3-319-84745-0   ISBN 978-3-319-51801-5   (eBook)
DOI 10.1007/978-3-319-51801-5

Printed on acid-free paper

This Springer imprint is published by Springer Nature
The registered company is Springer International Publishing AG
The registered company address is: Gewerbestrasse 11, 6330 Cham, Switzerland

Preface

Long-term digital preservation, the process of maintaining digital objects through time to ensure continued access, has become a crucial issue in recent years. The amount and the areas of digitized information are constantly increasing resulting in obsolescence of the software and hardware required to preserve digital information. Despite recognized need for preservation action, still more work is required to effectively address the issue in theory and practice.

The book is divided into two parts. Part I starts with discussion of the basic problems of long-term digital preservation. There are widely discussed and analyzed concepts and requirements for long-term digital preservation. Further, since metadata play important role in long-term digital preservation, processing of metadata in long-term digital archives is discussed. Long-term digital archives usually preserve all metadata transferred to them with corresponding digital objects and often use a subset of obtained metadata to manage archive assets. They also create preservation metadata, which describe processes and preservation actions applied to digital objects in the archives.

In Part II of the book, a framework based on the Digital Document Repository project CREDO is presented. Within the CREDO project, a demonstrative version of a digital repository enabling short- and long-term archiving of large volumes of digital resources has been designed and launched. The repository acts both as a secure file storage and as a digital archive providing metadata management and including the resources in archival packages. Reliability of information readouts is ensured by the repository through the data recording replication and monitoring mechanisms in the repository's file system, as well as through the distributed nature of the system that enables storing copies of the resources in more than one location. Advanced management system supports scheduling of operations on the archival storage while respecting the low energy consumption requirements.

One of the system primary functions is the support for various currently available data carriers such as hard drives, solid-state drives, and tapes. However, the repository architecture is multi-tiered and it enables (together with the emergence of new technologies) replacement and continuous upgrades of the individual components.

This solution has been designed for institutions that store large digital resources for long periods of time, e.g., cultural institutions, mass media, state administration offices, and healthcare institutions. An evaluation of our framework is presented, which illustrates the viability of our approach in retaining accessibility, authenticity, and usability.

Warsaw, Poland   Tomasz Traczyk
November 2016   Włodzimierz Ogryczak
                                                                                Piotr Pałka
                                                                        Tomasz Śliwiński

Acknowledgements

The project entitled *Digital Document Repository CREDO* was granted as a part of the pilot undertaking of the Polish National Centre for Research and Development, entitled '*DEMONSTRATOR+* Supporting scientific research and development works for demonstration scale' (Grant No. WND-DEM-1-385/00). The project was co-financed by the European Union through the European Regional Development Fund under the Operational Programme 'Innovative Economy' for the years 2007–2013, Priority Axis 1—Research and development of modern technologies.

Contents

Contributors

Włodzimierz Ogryczak Institute of Control and Computation Engineering, Warsaw University of Technology, Warsaw, Poland

Piotr Pałka Institute of Control and Computation Engineering, Warsaw University of Technology, Warsaw, Poland

Grzegorz Płoszajski Institute of Control and Computation Engineering, Warsaw University of Technology, Warsaw, Poland

Tomasz Traczyk Institute of Control and Computation Engineering, Warsaw University of Technology, Warsaw, Poland

Tomasz Śliwiński Institute of Control and Computation Engineering, Warsaw University of Technology, Warsaw, Poland

Part I
Problems of Long-Term Digital Preservation

Requirements for Digital Preservation

Tomasz Traczyk

Abstract Long-term preservation of digital information differs in many aspects from well-known short-term data storage. The longevity of the preservation causes a lot of little-known and difficult problems, which cannot be completely solved by technologies currently available, and some cannot be solved at all using only technical means. This chapter presents the most important requirements for long-term digital preservation and introduces some basic problems associated with this area.

1 Introduction

Long-term preservation of digital resources is proving difficult. The problem was not widely noticed until recently. Given the explosion of digital data, the issue takes on importance of the problem of our civilization. The costs of storing digital information are growing rapidly, and the risks increase.

The problem is of particular importance because of increasing prevalence of 'born digital' objects, which have no other representation than digital, so they are seriously threatened with the total loss.

In February 2015 Vint Cerf, one of the "fathers of the Internet" and a Google vice-president, warned against 'Digital Dark Age', when future generations will have little or no record of the 21st Century, as the digital data will be lost or unreadable as necessary hardware or software becomes obsolete [3].

Since long-term preservation of digital information differs in many aspects from well-known short-term data storage, requirements for this type of preservation should be clearly formulated, and the differences highlighted.

T. Traczyk (✉)
Institute of Control and Computation Engineering,
Warsaw University of Technology, Warsaw, Poland
e-mail: T.Traczyk@ia.pw.edu.pl

© Springer International Publishing AG 2017

T. Traczyk et al. (eds.), *Digital Preservation: Putting It to Work*,
Studies in Computational Intelligence 700, DOI 10.1007/978-3-319-51801-5_1

2 The Concept of Digital Preservation

The goal of digital preservation is to keep digital materials not only technically accessible, but also usable for long periods of time.

Though the digital information can be losslessly reproduced, it is generally more vulnerable to the loss than its analog counterparts. Media degradation and technological obsolescence caused by rapidly emerging new technologies are the main technical threats. Difficulty in understanding stored digital content, inability to determine what represents the given object and in what context it was created, and impossibility of finding the desired resources in the ocean of digital data are even more dangerous threads, as they cannot be solved by purely technical means.

More detail objectives of digital preservation are:

- reliable storage of information;
- long-term protection and authorized use of the stored resources;
- usability of the information in the distant future, which means at least the possibility of correct reading from data carriers and proper interpretation of data formats;
- the ability to locate the sought resource and restore the context of its creation;
- verifiability of integrity and authenticity, and non-repudiation of stored resources.

2.1 Long-Term Preservation

Long-term preservation means ensuring continued access to digital materials, or at least to the information contained in them, for a period of dozens or hundreds of years, or even indefinitely. Therefore, the preservation has to extend beyond the foreseeable future and beyond the life time of any digital technology.

Long-term preservation turns out to be much more difficult problem than short-term storage since:

- the preservation period extends beyond life time of existing data carriers, which means, that the data will have to be transferred to new media, probably many times;
- the preservation extends also beyond use period of file formats, which means, that the data probably will have to be migrated to new formats, perhaps many times;
- data transfers and, in particular, format migrations introduce the risk of distortion of the information;
- the longevity of the preservation generates also non-technical risks, e.g.

  - cessation of existence of the organizations, which owns the rights to the information,
  - decommissioning of the institution, which is responsible for the preservation,
  - disappearance of so-called designated community [2], i.e. potential information consumers, who are able to understand and utilize its content;

- long period of the preservation causes considerable costs.

2.1.1 Deep Digital Archive Concept

Usually the main task of an archive is to preserve the holdings, rather than sharing them. In case of a digital archive, this fact is of great importance, because it significantly affects the technological assumptions.

If the stored resources need not to be accessible 'on request' (on-line), but are delivered rarely, 'on order', i.e. with acceptable delay of several hours or even days, we are dealing with a so-called deep archive.

Storage hardware used for digital deep archives can usually be less expensive then used in on-line systems. Moreover, a deep archive can store the archival resources in such a kind of storage, which needs not to be permanently on-line. It means, that devices like tape libraries or optical disk jukeboxes can be used with no performance problems. In case of disk-based storage, the disks and supporting servers can be powered-off most of the time, and must be switched on only when necessary, e.g. when the archival data are to be transferred or some administrative tasks are to be performed. It is very important, since the electric energy costs are a substantial part of the digital archive operating costs.

Such an archive is able to meet the requirements of cost efficiency. The efficient long-term digital archive should, therefore, be designed and should operate as a deep archive.

2.2 Bitstream Versus Content Preservation

Bitstream preservation means ensuring faithful storing digital data 'bit by bit'. The correctness of this kind of preservation is fairly easy to verify, e.g. with use of digital digests.

Content preservation stands for maintaining long-term usefulness of the stored information, which requires the ability to correctly interpret the stored data. This requirement is usually difficult to verify. In many cases the verification by technical means is impossible at all, and an evaluation by a human consumer is necessary.

Unfortunately, correct bitstream preservation of a digital resource does not automatically cause the durability and usability of its content. Though the stream of bytes may be restored without any distortions, we may not be able to properly interpret it, e.g. the format of the data may be unknown or the software necessary to 'decrypt' the format may become unavailable or cannot be run in available hardware/system environments.

Even if it is possible to restore the resource, and to interpret formats of its files, the restored data may be useless if no consumer exists, who can understand the information. It means, that to preserve the usability of the archived content, existing designated community is necessary. Therefore, the content preservation cannot be realized by only technical means.

On the other hand, a perfect bitstream preservation is not always necessary to achieve proper content preservation. Since the usefulness of the preserved content often depends on the adequate interpretation by a human, some distortions in bit-stream may be acceptable or even irrelevant; e.g. if single pixels on a photography or on random film frames are distorted, the whole material is still usable and may be practically indistinguishable from undistorted one.

Though such an 'imperfect bitstream preservation' may be sufficient in many cases, it is very difficult to asses whether the given distorted object is faithful enough. Therefore, in practical archival systems a perfect bitstream preservation is usually required. Some distortions may only be tolerated in case of resource reconstruction after its partial destruction, but this is usually a process controlled by a human.

An ultimate goal of long-term digital archiving is to ensure content preservation. Since this kind of preservation cannot be achieved only by appropriate technical solutions, it is usually only partially guaranteed by the archive itself, and must be complemented by some external activities, e.g. taken by appropriate designed communities.

3 Requirements for Long-Term Digital Preservation

Digital preservation can be defined as long-term if it meets many requirements concerning information longevity, verifiability, availability, etc. (cf. [4, 11]). The most important requirements are discussed in this section.

3.1 Information Longevity

Information longevity means at least its long-term readability in the sense of accurate preservation of bits (bitstream preservation, see Sect. 2.2). Unfortunately, due to the lack of sufficiently reliable media, even this kind of longevity is currently difficult to guarantee, as discussed in Sect. 4.

In a broader sense, the longevity should mean an ability to utilize the information after many years. Therefore, at least hardware and/or software means have to be available on long-term horizon to read and interpret the data, e.g. to 'decrypt' its format.

As the digital data must be subject to various operations to keep that sense of longevity, it is necessary to enable management of archived data resources. Due to rapid progress in computing and storage technologies, if the information is to be permanent, it has to be able to be transferred to new data carriers and migrated to new formats. Thus, used technologies and formats must ensure feasibility of these operations and some possibility of verification of the results.

3.2 Verifiability of the Information

Verifiability of the stored information correctness is crucial to ensure the information persistence. Information integrity and authenticity are subject to the verification.

In some cases, a non-repudiation of the stored resources must also be assured. This usually cannot be achieved by technical-only means, as it needs appropriate contract between the archive and the information producer, and proper technical and/or non-technical procedures to confirm the correctness and origin of the stored resource by its producer. If the origin of the resource is credibly confirmed, and its authenticity and integrity are preserved, the resource can be recognized as non-repudiable.

The archived documents should also be portable in this sense, that after their transfer to another archive the ability to verify their integrity and authenticity should be maintained [4].

3.2.1 Information Integrity

Information integrity means its completeness and conformity of the content with its declaration (also of the data formats used).

In order to check the integrity, the metadata for each archived resource must be delivered, which declares the resource content, or at least the initial/original state of the resource must be captured, confirmed by the producer, and stored. Preserving the metadata in the archive is therefore necessary for the possibility to verify the integrity.

3.2.2 Information Authenticity

Information authenticity may have very sophisticated interpretations [8]. In the simplest sense it means that no damage or unauthorized change of the information happened during its ingest and preservation. This means a need for something more than integrity: detailed verification of the information immutability is necessary. If, however, a resource changes in any way, a detailed trail of each change, together with identification of its originator and his permissions must be recorded.

In digital archives the data immutability is usually checked with use of digital digests. Tracking history of resource changes needs creating preservation metadata and storing them together with the archived resource.

3.3 Availability of the Information

Availability of the information requires its easy finding, effective obtaining and ability of its proper interpretation.

The stored data must be permanently accessible to authorized consumers, independently of possible storage failures. We must be able to obtain undistorted resources. Therefore, the archive system must be highly reliable and probably involve some redundancy.

Moreover, we must be able to properly interpret the content of the resource even after very long time from its origin. This feature is difficult to maintain because of the "moral" aging of file formats. Therefore, not all popular file formats are suitable for long-term archiving.

3.3.1 File Formats and Preservation

Since the stored information needs to be properly interpreted after years, file formats used in a digital archive have to provide significant probability of correct data read in the distant future, if possible without the use of dedicated or proprietary software.

File formats vary in usefulness for long-term storage. The Library of Congress proposes [7] the following criteria for the formats evaluation:

- disclosure—available and complete format specification, and availability of tools for checking the technical correctness of the file format;
- adoption—popularity of the format, and availability of format-related tools;
- transparency—the ability to analyze the data without specialized tools;
- self-documentation—metadata contained within the files;
- external dependencies—dependence on specific hardware and software environment;
- impact of patents—the degree of limitation of long term preservation by patents;
- technical protection mechanisms—solutions for future migrations, changes in media and legal rights protection.

Some additional criteria are formulated by The National Archives of Great Britain [1]:

- stability—format specification should not be subject to frequent changes, and new versions should ensure compatibility with previous versions;
- re-usability—ease of re-use.

Generally, proprietary formats are less suitable for long-term preservation than open ones, and technically complex formats are worse than simpler ones. The most appropriate are probably text-based formats, e.g. utilizing XML or similar standards.

Using these or similar criteria, the archive should qualify specific formats as recommended or acceptable. It is desirable that only recommended or acceptable formats are allowed to be stored in the long-term archive.

3.4 Confidentiality of the Information

Any information stored in the archive has to be protected against unauthorized access. If any legal requirements apply to the preserved resources, they must be satisfied.

3.5 Economical Effectiveness of the Storage

Long-term preservation is inherently expensive, as it requires providing many services for a very long time. The costs must be reasonable, however, to prevent discontinuation of resources preservation due to the exhaustion of funds.

One of the major cost components of digital preservation, especially if disk-based storage is used, is the cost of electric energy. Therefore, the storage must be designed and managed in a way that ensures energy efficiency.

3.6 Standardization of Digital Archives

Since a long-term archive has to be exploited and maintained for several dozens of years, and archived resources should be possible to read and interpret for even much longer time, the design and implementation of the archive must be based on standardization. Only a compliance with standards can assure, that the system itself, and the stored files, will be understood and properly interpreted by technical means and human users in distinct future.

Reference Model for an Open Archival Information System (OAIS) [2], prepared by Consultative Committee for Space Data Systems (CCSDS), and accepted by International Standard Organization as ISO 14721:2012 standard [6], appears to be the most important standard for digital preservation. The standard defines basic concepts and reference model of the archive, and model structure of archival object (so-called information package). The reference model defines proper terminology and recommends, among others: operating environment for the archive, the division of responsibility, functional model of the archive and data flows, general data models for the archive, general API models, patterns of information migration, rules of cooperation between archives and general layered model of archive software.

File formats accepted by the archive should be subject to the standardization. Since metadata of the preserved resources play a crucial role in preservation, verification and proper interpretation of the resources, the metadata themselves should also be standardized, as well as file formats used to store the metadata. Extensive information about the standards and formats for metadata can be found in chapter "Metadata in Long-Term Digital Preservation".

Since the resources stored in the archive are to be read in distant future, when current technologies may be long forgotten, it is advisable to preserve documentation of used technologies: media and file formats, metadata standards and formats, etc., in the archive, possibly in the most readable form, e.g. as simple text files.

3.7 Certification of Digital Archives

The digital archives should prove to be trustworthy. Three main factors should be examined to consider an archive worthy of the term:

- the organization of the archive: processes of data receipt, storage and transfer to customers;
- correctness of applied procedures and reliability of the data storage;
- the credibility of the institutions that carry out evaluations of the archives.

To be recognized as trustworthy by potential customers, digital archives need certification. Special attention should be paid on the transparency of the procedures used in the archives, and transparency of ratings, which archives gain in audits and certification processes.

The document "Trustworthy Repositories Audit & Certification: Criteria and Checklist" (TRAC) [9] uses the concepts contained in the OAIS model. The document supports auditing of archives, and making self-esteems. Measures used by TRAC to assess the reliability of the repository are divided into three groups:

- organizational infrastructure (functions, financial, and legal administrative, human resources);
- management of digital objects;
- technology, technical infrastructure and security.

Currently the main standard used for certification of digital archives is "Trusted Digital Repository", approved as ISO standard ISO 16363:2012 [5]. The standard defines criteria for the certification of digital archives and includes so-called TDR-checklist ("Trusted Digital Repository Checklist").

4 Means for Digital Preservation

4.1 Storing Data on Reliable Devices

It can be expected that the accurate[1] and long-term preservation of the data might be ensured by data carriers, which guarantee long durability time. Unfortunately, so far no durable enough digital data carriers exist.

---

[1]In this section we assume, that perfect bitstream preservation is required by the archive (see Sect. 2.2).

Records on the most popular magnetic data carriers—tapes and disks—have durability of only several years, because of magnetic particle instabilities [10]. To prevent data loss, the signal recorded on magnetic carriers must be refreshed from time to time, which is particularly troublesome in case of tapes. The carriers themselves have also limited time of life and should be replaced after no more than several years. Disk systems (e.g. disk arrays) are expensive and energy-intensive, and they are less reliable than expected. Storage based on flash memory and SSD disks is considered fairly durable, but unfortunately is still too expensive.

Optical discs have persistence limited to single dozens of years, low capacity, slow record, they are cumbersome to use and susceptible to damage. Some new types of optical disks have declared life for hundreds of years, but they are very costly, and the declarations of their durability are called into question.

The reliable data preservation on magnetic tapes and optical disks needs proper storage conditions, as temperature and humidity. If the conditions deviate from the recommendations, the durability of the records is greatly reduced.

Moreover, most currently used data carriers are not immune to electromagnetic pulse that may occur, for example, as a result of increased solar activity. Some novel methods of data storage, e.g. digital record on film tape, are resistant to this threat, but are not very capacious and poorly tested.

Since digital technologies evolve very quickly, even commonly used types of media quickly become obsolete, and after a few years it may be impossible to find a working drive, which is able to read the media. This problem applies particularly to magnetic tapes.

As a result, 'analog' originals, e.g. classic film tapes, are still more durable than their digital copies. Storage of non-digital originals is usually also less expensive then preservation of their digital equivalents. Unfortunately, more and more resources are 'born digital', so they have not non-digital versions.

4.2 Preservation by Copying

Since the reliability of storage on existing data carriers is inadequate, it is necessary to replicate the information to many carriers. The resources should also be periodically relocated to new data carriers, perhaps of a novel type.

If replicas are dislocated, i.e. stored in geographically distant locations, the replication prevents also the impact of natural disasters, sabotage, etc.

Additional protection, which increase the probability of correct read and interpretation of resources, is to store their replicas in diverse technological environments and different recording formats.

4.3 Metadata in Preservation

Preserving metadata together with the archived resources is necessary for many crucial archive features, e.g. verification and proper interpretation of the resources.

As the metadata creation, preservation, standardization and utilization are important, but complex issues, they are widely discussed in the next chapter.

5 Conclusion

Though the long-term digital preservation causes problems, which we currently are not able to completely solve, a designer of a digital archive should take all the known problems into consideration, as some of them will probably become solvable in more or less distant future.

The design of the archive should enable adopting new technologies and methods. If the complete requirements (including widely recognized standards), and potential issues and threads are known during the design process, the resultant system may be more robust and flexible, ensuring longevity of the archived information, as well as a small risk and reasonable costs of future system modernizations.

References

1. Brown, A.: Digital preservation guidance note 1: Selecting file formats for long-term preservation (2008). URL http://www.nationalarchives.gov.uk/documents/selecting-file-formats.pdf. Access: 2016-10-25.
2. Consultative Committee for Space Data Systems: Reference model for an open archival information system (OAIS). Recommended practice. (2012). URL http://public.ccsds.org/pubs/650x0m2.pdf. Access: 2016-10-25.
3. Ghosh, P.: Google's Vint Cerf warns of 'digital Dark Age'. URL http://www.bbc.com/news/science-environment-31450389. BBC News. Access: 2016-10-25.
4. Huhnlein, D., Korte, U., Langer, L., Wiesmaier, A.: A comprehensive reference architecture for trustworthy long-term archiving of sensitive data. In: 2009 3rd International Conference on New Technologies, Mobility and Security, pp. 1–5. IEEE (2009).
5. International Standard Organization: Space data and information transfer systems – audit and certification of trustworthy digital repositories (2012). URL http://www.iso.org/iso/catalogue_detail.htm?csnumber=56510.
6. International Standard Organization: Space data and information transfer systems – open archival information system (OAIS) – reference model (2012). URL http://www.iso.org/iso/catalogue_detail.htm?csnumber=57284.
7. Library of Congress: Sustainability of digital formats (2013). URL http://www.digitalpreservation.gov/formats. Access: 2016-10-25.
8. Lynch, C.: Authenticity and integrity in the digital environment: an exploratory analysis of the central role of trust. Tech. rep., Council on Library and Information Resources, Washington, D.C. (2000).
9. OCLC, CRL, NARA: Trustworthy repositories audit & certification: Criteria and checklist (2007).

10. Van Bogart, J.W.: Magnetic Tape Storage and Handling: A Guide for Libraries and Archives. Commission on Preservation and Access and the National Media Laboratory (1995). ISBN 1-887334-40-8.
11. Wallace, C., Pordesch, U., Brandner, R.: Long-term archive service requirements (2007). URL http://www.ietf.org/rfc/rfc4810.txt. Access: 2015-12-01.

Metadata in Long-Term Digital Preservation

Grzegorz Płoszajski

Abstract Metadata has been used to describe digital data for several years. First metadata standards were created in the nineties. Particularly important proved to be Dublin Core designed for the description of web-based resources (1995). At the same time Exif standard was created, determining mainly technical metadata used to describe parameters of digital images and information on equipment used. Since that time various other types of metadata have been created, e.g. rights metadata, provenance, structural, and many domain-oriented standards. Long-term digital archives usually preserve all metadata transferred to them with corresponding digital objects and often use a subset of obtained metadata to manage archive assets. They also create preservation metadata, which describe processes and preservation actions applied to digital objects in the archives. In this chapter basic information about metadata in long-term digital archives is given.

1 Introduction

1.1 Digital Preservation

The goal of digital preservation is to keep digital materials accessible and usable for long periods of time. Digital information is fragile. It can be lost because of media degradation. Another problem is technological obsolescence caused by introducing new software and new technologies, which supersede older. The older fall out of use.

Digital preservation must be performed actively. Preservation activities can be applied to stored digital information in institutional repositories, digital libraries and archives; hereinafter referred to collectively as digital archives. Archives apply such strategies as data redundancy, migration and emulation. Data redundancy (making several copies of digital objects) may be helpful in short-term. Migration can be done to new formats of files or to new media; it solves to some extent problem

G. Płoszajski (✉)
Institute of Control and Computation Engineering, Warsaw University of Technology,
Warsaw, Poland
e-mail: G.Ploszajski@ia.pw.edu.pl

© Springer International Publishing AG 2017   15
T. Traczyk et al. (eds.), *Digital Preservation: Putting It to Work*,
Studies in Computational Intelligence 700, DOI 10.1007/978-3-319-51801-5_2

of technological obsolescence. Migration is addressed to long-term preservation. Emulation can help when new software supersedes older; files do not have to be migrated to new formats. But emulation software must be updated.

1.2 Preservation and Metadata

Generally it is a good idea to think about preserving digital objects when they are created. One of possible activities is gathering as much information about each digital object as possible from the start. Such information is called metadata.

Common practice in digitization of cultural heritage is creating metadata right after creating corresponding digital objects. In such case metadata should contain information about original object and about the created digital one (e.g. parameters of digital image, equipment used, data and time of creation etc.). Metadata can contain supplementary information concerning quality check of created object (e.g. description of flaw on an image) or process of additional transformations of object (e.g. elimination of skew). It should contain legal information concerning access to digital object and allowed, and/or forbidden transformations. Metadata is called according to nature of information, e.g. descriptive, technical, administrative, rights, provenance.

Many digital objects are created for current utility purposes; the need to preserve these objects may become conscious after some time. Personal collections of photo images might be an example. Creating metadata in such case is more difficult and for some types of metadata even impossible.

When digital objects together with corresponding metadata are 'sent' to digital archive to be preserved for a long time, process of preservation starts. Another type of metadata is used to keep record of preservation activities made in the archive. This group is called preservation metadata. An archive attempting to preserve digital data for long time should use preservation metadata. When digital object is returned to the owner after a long time it should be accompanied by the metadata, which will inform the user of its content.

Systems of metadata are usually defined in a formal way. They are often called standards (some even passed long way to become ISO standards). Most of metadata systems use XML notation. XML Schema can be treated like a simple standard definition. Some systems of metadata are described with use of dictionaries. One of first standards of preservation metadata is PREMIS, managed by the Library of Congress.

1.3 Short and Long-Term Preservation

Digital Preservation Handbook Glossary [1] developed by Digital Preservation Coalition defines short, medium and long-term preservation as follows:

- *Short-term preservation*—Access to digital materials either for a defined period of time while use is predicted but which does not extend beyond the foreseeable future and/or until it becomes inaccessible because of changes in technology.
- *Medium-term preservation*—Continued access to digital materials beyond changes in technology for a defined period of time but not indefinitely.
- *Long-term preservation*—Continued access to digital materials, or at least to the information contained in them, indefinitely.

In practice terms "short-term" and "medium-term" are met rather seldom, however the case where information data should be preserved only until some defined period of time can happen often, usually when there is a legal obligation to store some kind of documents (e.g. master thesis in Poland should be preserved during 50 years).

There is also another group of information which should be stored to guarantee usability of digital data (content and metadata) after long time but is not treated as metadata. This can be standards of metadata and dictionaries of terms used in metadata. Without them after a long time information contained in metadata might be difficult to understand.

2 Metadata in Digital Archives

2.1 Digital Archive and Customers

For the purpose of this chapter we consider model of digital archiving containing two elements:

- digital archive *and*
- customers.

Customers entrust their digital data to the archive and archive takes responsibility for keeping these data usable for long time, and provide access to the data for the customers. This may be university repository and scientific staff of a university or business oriented digital archive offering services and external customers, individual or companies.

The data can be transported to the archives either by network or on physical media, e.g. hard disks or CDs. In the second case the archive reads and store the data from the media. In both cases there should be a procedure according to which customer can make sure that digital data in the archive is an exact copy of the original.

In such model it is usually the customer's responsibility to provide metadata; the data sent to the archive should consist of the 'content' data (e.g. digital images, sound, text, measurement data) and metadata. Archive becomes responsible only for keeping metadata sent by customers and creation of preservation metadata.

Responsibility of the archive and the customer concerning the 'content' data and metadata can be formally based on agreements between them. Agreements may state accepted formats of files and standards of metadata as well as procedures of delivery of data to the archive (SIP) and of sending it back to the user after a "long" time (DIP).

Metadata is often external to the content file—saved in separate file—however can also be embedded in the content files, especially technical metadata.

The model consisting of archive and customers does not fit to web archiving, where the content of the web is often grabbed as it is. Creators of the web content are neither customers nor partners of the archive. They do not have to adhere to metadata standards. They can use different standards. They do not have to create metadata at all. The archive must manage such inconsistent information. We shall not consider such model in this chapter.

2.2 Metadata Received and Metadata Created in the Archive

In the context of processing data in long-term archives metadata can be divided into two groups:

- metadata supplied to the archive together with related digital objects,
- metadata created in the archive.

There are various types of metadata supplied to archives, e.g. descriptive metadata, technical, structural, administrative, rights, provenance, business [2]. Technical metadata can contain parameters of digital image or audio file. Structural metadata can show how a complex object is built and what is a role of individual data files. Structural metadata can also represent structure of SIP packages sent to the archive.

Some of the above mentioned metadata are created manually, especially description metadata, some are created by hardware, e.g. technical metadata of images in photo cameras, and some by software, e.g. measurement metadata of scientific experiments or space missions.

Second group contains preservation metadata, which make record of operations made in the archive, e.g. transformation of SIP to AIP, counting checksums, comparing checksums of different copies of individual files, migration to new file formats or migration to new memory media. These metadata should be created by software used for management of archived data.

However some metadata from first group can also be included in preservation metadata. Digital objects can be created as 'born-digital' or as digital 'copies' of physical (analogue) objects, e.g. images of paper documents or museum artifacts. Such objects can be processed, e.g. to eliminate skew or noise or simply to change format (e.g. RAW to TIFF, Word to PDF/A). History of such operations should be recorded in provenance metadata. Provenance information may be important in the future for assessing authenticity of digital objects and to make judgments about whether information is trusted. (Term *provenance* is discussed in detail in [3].)

Generally, provenance metadata should be gathered by the customers from the time of creating the object and transferred to the archive; provenance information can be included in preservation metadata during ingest and supplemented during preservation activities.

Metadata sent to archive can be stored separately from the content files, however in long-term archives metadata is usually treated as part of digital object; all metadata can be subject to preservation activities together with related content files.

When digital object comes to an archive, part of its metadata is often copied to a database (or other data structure) to help in management of archive's assets. Part of metadata transferred to management database should contain:

- all or some of descriptive metadata to help in searching objects,
- a small number of technical metadata to help make decisions concerning migration to new formats,
- rights metadata to control access to stored objects or to information on these objects.

2.3   Evolution of Metadata Standards

2.3.1   New Versions of Standards

Metadata standards evolve. They get new versions, e.g. standard VRA Core [4], used for the description of images and work of art and culture, created in the late nineties, got version 4.0 in 2007; ISO 19115 standard DIF [5] used in earth sciences, got version 6.0 in 2010; METS [6], created in 2001, got version 11.0 of XML Schema in 2015. As a result metadata based on different versions of metadata standards must co-exist in long-term digital archives and archives must be able to manage assets containing metadata based on the variety of standard versions.

A number of data dictionaries and ontologies have been developed as a part of metadata standards to help in preparing consistent metadata (e.g. PREMIS [7]). New standard versions get new dictionaries and ontologies.

Archives should store metadata standards with these dictionaries and ontologies (as well as taxonomies and thesauri). After many years they should help to understand what is the real meaning of some metadata. They might also help in precise searching.

2.3.2   Development of Dublin Core

Dublin Core is domain-agnostic standard, created during series of workshops which began in 1995 in Dublin, Ohio. Simple Dublin Core Metadata Element Set(DCMES) has 15 elements: Title, Creator, Subject, Description, Publisher, Contributor, Date, Type, Format, Identifier, Source, Language, Relation, Coverage, Rights. Each element is optional and repeatable. Descriptive names 'promote' a common semantic understanding of the elements. List was published in 1998 [8]. Using controlled vocabularies for specific elements could increase interoperability of the standard.

A number of domain-oriented metadata standards was inspired by this list of elements, e.g. EBU Core metadata set for Radio archives, Tech 3293, version 1.0, 2001 [9] (EBU—European Broadcasting Union [10]). Some standards got mappings from Dublin Core, e.g. Directory Interchange Format DIF used in Earth Science [11], or to Dublin Core.

Another achievement was creation by *Open Archives Initiative* Protocol for Metadata Harvesting (OAI-PMH) [12], based on obligatory use of Simple Dublin Core.

Dublin Core has been used with types of materials which demanded some complexity, e.g. linking dates with events or persons with functions. As result of discussions qualifiers have been developed. Since then the two versions of Dublin Core were distinguished: the older one called *simple* or *unqualified* and the new one called *qualified* with qualifiers [13].

Qualifiers could be used to refine an element. For example element *Date* got five refinements: *Created, Valid, Available, Issued, Modified* and element *Relation* twelve refinements: *Is Version Of, Has Version, Is Replaced By, Replaces, Is Required By, Requires, Is Part Of, Has Part, Is Referenced By, References, Is Format Of, Has Format*. Qualifiers could also be used to indicate encoding scheme. For example element *Subject* got five controlled vocabularies (i.a. LCSH, MeSH, UDC), element *Language*—two standards: ISO 639-2 abd RFC 1766.

Qualifiers did not solve all problems. Standard was changed. New version is called DCMI Metadata Terms [14]. 55 elements have been defined: (*abstract, accessRights, accrualMethod, accrualPeriodicity, accrualPolicy, alternative, audience, available, bibliographicCitation, conformsTo, contributor, coverage, created, creator, date, dateAccepted, dateCopyrighted, dateSubmitted, description, educationLevel, extent, format, hasFormat, hasPart, hasVersion, identifier, instructionalMethod, isFormatOf, isPartOf, isReferencedBy, isReplacedBy, isRequiredBy, issued, isVersionOf, language, license, mediator, medium, modified, provenance, publisher, references, relation, replaces, requires, rights, rightsHolder, source, spatial, subject, tableOfContents, temporal, title, type, valid*). In the dictionary every element has URI identifier in corresponding namespace using persistent uniform resource locators (http://purl.org/dc/terms/).

Dublin Core was published as ISO Standard 15836 in 2009. Not all problems of users, which inspired changes, have been resolved by this version. Ability to link persons to their roles was improved, nevertheless was not satisfactory for many users. Standard DCMI Metadata Terms allows to link agents to roles from closed list, being a small subset (about 30 elements from 150) of *MARC Code List for Relators* [15].

Dublin Core was defined without using XML notation. However there are published guidelines for implementing Dublin Core in XML [16]), and validating with XML Schemas: for Simple Dublin Core [17] and for DC Terms [18].

There is a number of digital libraries which use Dublin Core as descriptive metadata. dLibra software, which is popular in Poland, allows to use simultaneously Simple Dublin Core and other descriptive metadata standards (using Simple Dublin Core is obligatory). OAI-PMH Protocol (*Open Archives Initiative Protocol for Metadata Harvesting* [12] allows to harvest metadata from libraries or archives, which also use this protocol. Some large archives make use of protocol OAI-PMH, i.a. arXiv [19]. In Poland metadata from all dLibra libraries and some others using OAI-PMH, are harvested to the central library of Digital Libraries Federation FBC [20] and made available for searching.

The above mentioned OAI-PMH protocol has created own namespace and own XML Schema, based on DCMI schema for unqualified Dublin Core.

2.3.3   Development of Domain-Oriented Metadata Standards

Domain oriented metadata standards have been subject to similar development as Dublin Core and changed a lot. However motivation to making changes usually was different.

One example is EBU Core, which in 2001 in first version with its 15 elements was almost identical to Simple Dublin Core [9]. In 2015 version 1.6 of standard was constructed in totally different way [21]. Changes introduced in version 1.6 are explained in introduction to specification in the following way: "EBUCore 1.6 takes into account latest developments in the Semantic Web and Linked Open Data communities. EBUCore 1.6 is available as a RDF ontology entirely compatible with the W3C Media Annotation Working Group ontology, which model is common and based on the EBU Class Conceptual Data Model (Tech.3351). A link to the RDF/OWL ontology and its documentation is provided in Annex B. The EBUCore ontology has been updated to complement EBU's CCDM (Tech 3351) and improve mapping with other ontologies. EBUCore RDF is listed as Linked Open Vocabulary as well as RDF-Vocab for Ruby developers. The definitions in EBUCore 1.6 have been refined. The schema structure has been reinforced for registration in EBU's Class 13 in SMPTE. The new advanced data model for audio defined in Tech 3364 and introduced in EBUCore 1.5 has been updated to reflect discussions around its adoption in ITU." Thus creation of version 1.6 was motivated not only by 'internal' EBU problems but also to enhance business cooperation of European producers with US (SMPTE—Society of Motion Picture and Television Engineers [22] is an American organization setting standards for Motion Imaging) and to be in touch with developments taking place in International Telecommunication Union [23] (ITU is the United Nations specialized agency for information and communication technologies ICTs).

Interesting history have Photo metadata standards created by International Press Telecommunications Council (IPTC) [24] to improve exchange of news among newspapers and agencies (see Metadata History Timeline and Guidelines). Metadata can be embedded in photo files. IPTC Photo Metadata standard is supported by such software as Adobe Photoshop.

Great role of Simple Dublin Core remains also valid in version DCMI Metadata Terms.

2.4   Minimal Set of Metadata

Archives can define requirements concerning data and metadata delivered to them. Sometimes this is done in form of defining minimal requirements.

American National Archives and Record Administration (NARA) applied such approach to metadata. NARA published in 2015 *Metadata Guidance for the Transfer of Permanent Electronic Records* [25].

Recommendation is addressed to federal agencies. They must prepare metadata which accompany transfers of permanent electronic records to national archives. Besides descriptive, technical and administrative information concerning structure and content of electronic records "metadata elements also provide contextual information that explains how electronic records were created, used, managed, and maintained prior to their transfer to NARA, and how they are related to other records. This information enables NARA to appropriately manage, preserve, and provide access to electronic records for as long as they are needed."

NARA defined minimum metadata requirements as a subset of the Dublin Core Metadata Element Set v.1.1.

"Agencies should provide the following elements for each file or item included in a transfer (…):

1. Identifier [File Name]. The complete name of the computer file including its extension (if present);
2. Identifier [Record ID]. The unique identifier assigned by an agency or a records management system;
3. Title. The name given to the record;
4. Description. A narrative description of the content of the record, including abstracts for document-like objects or content descriptions for audio or video records;
5. Creator. The agent primarily responsible for the creation of the record;
6. Creation Date. The date that the file met the definition of a Federal record; and
7. Rights. Information about any rights or restrictions held in and over the record including access rights such as national security classification, or personally identifiable information, Privacy Act, or Freedom of Information Act, or usage rights relating to copyright or trademark.

Agencies should provide the following metadata elements, if they apply to the record being transferred:

1. Coverage. The geographic and temporal extent or scope of the content of the record; and
2. Relation. The relation element should be used if a record is composed of multiple files that form a logical record, or is a necessary component of another logical record.

If an agency provides additional metadata elements, NARA will accept that metadata as part of the transfer process in addition to NARA minimum metadata requirements. Agencies should notify NARA of any metadata standards that are in use with permanent electronic records and provide relevant schemas, data dictionaries, controlled vocabularies, ontologies, and system indexes at the time of transfer."

For the element Rights five refinements were given (four new; only RightHolder was defined in DCMI Metadata Terms):

- Security Classification—mandatory,
- Previous Security Classification—mandatory if applicable,
- Access Rights—mandatory,
- Usage Rights—mandatory when applicable,
- Rights Holder—mandatory when applicable.

2.5 Comments on File Formats

Some repositories are oriented to achieve two goals: provide short-term access and long-term usability. To imagine what "short time" can be, consider problems one might have with reading text documents prepared with word processing programs 25 years ago. Certain file formats might become not accessible for current applications. To read texts from this archive an extra software may be needed. In case of short-term archives it is rather the users' problem to find and use such software, while long-term archives should convert files, which might become obsolete to new formats. Such operation is called format migration in digital archives.

Migration is rather costly. Archives try to limit number of file formats which they accept.

Analysis of digital formats with respect to their usability in long-term digital preservation have been made for long time by many organizations. The Library of Congress manages the gathered results [26] using title *Sustainability of Digital Formats Planning for Library of Congress Collections*. There is also given comprehensive information concerning evaluation factors taken into account, such as:

- Sustainability factors—apply for all categories of information:

  - Disclosure,
  - Adoption,
  - Transparency,
  - Self-documentation,
  - External Dependencies,
  - Impact of Patents,
  - Technical Protection Mechanisms;

- Quality and functionality factors (for selected content types):

  - Still Images:
    Clarity (support for high image resolution),
    Color maintenance (support for color management),
    Support for graphic effects and typography,
    Support for multispectral bands;

– Sound:
 Fidelity (support for high audio resolution),
 Support for multiple channels (including note-based, e.g., MIDI),
 Support for downloadable or user-defined sounds, samples, and patches;
– Text:
 Support for integrity of document structure and navigation,
 Support for integrity of layout, font, and other design features,
 Support for rendering for mathematics, formulas, diagrams, etc.;
– Moving Images:
 Clarity (support for high image resolution),
 Fidelity (support for high audio resolution),
 Support for multiple sound channels.

Some archives make lists of preferred file formats or of accepted file formats. Good example of such practice is shown in a *Revised Format Guidance for the Transfer of Permanent Electronic Records* [27]. American National Archives and Record Administration (NARA) defined in this document ten categories and a few subcategories of digital resources:

1. Computer Aided Design,
2. Digital Audio,
3. Digital Moving Images:

   a. Digital Cinema,
   b. Digital Video,

4. Digital Still Images:

   a. Digital Photographs,
   b. Scanned Text,
   c. Digital Posters,

5. Geospatial Formats,
6. Presentation Formats,
7. Textual Data (plain text, formatted text, word processed),
8. Structured Data Formats (databases, spreadsheets, scientific data),
9. Email,
10. Web Records.

Seven categories are common with the Library of Congress list [26] (category *Datasets* can be treated as equivalent to *Structured Data Formats*). Three: *Computer Aided Design, Presentation Formats* and *Email* have no equivalent categories. The Library of Congress defines also eighth category *Generic*, designed for wrappers (e.g. RIFF), bundling formats (which can be used to construct SIPs and DIPs) and encodings.

For each of ten categories NARA declared formats using up to three levels of format acceptance:

- Preferred Formats,
- Acceptable Formats,
- Acceptable for Imminent Transfer Formats.

The third level is designed for "legacy formats that are no longer in common use and that NARA will eventually stop accepting". Special procedure of transfer of such files is required. In most cases only two initial categories of formats are defined. The third one is given only for two categories: 5 and 8.

Decisions made by American archives can be inspiring for other archives.

Migration of file formats can have implications for metadata. If metadata is external, the information on change of content file format can be stored in preservation metadata and in Data Management; there is no need to change original metadata. In case of embedded metadata, migration of content to new format may lead to problems if new format is unable to embed original metadata. However, every archive which accepts files with embedded metadata should be prepared to preserve metadata. At least it can export embedded metadata to external file at the time of ingest.

3   Metadata of Typical Digital Archive Resources

3.1   Categories of Digital Resources

Digital archives gather data coming from various domains of life. Description of those data is usually domain-oriented and there are numerous standards of descriptive metadata. Categories of digital objects are less diverse, because they can be used in many domains. For example still images can be used to make digital copies of analogue originals such as paper documents and museum artifacts, and to create and save "born digital" objects such as medical documentation or satellite photographs.

Problems concerning long-term preservation of digital images are similar regardless of domain of the application. In Sect. 2.5 are shown ten categories of digital objects, formulated by the American National Archives [27]. Categories listed by the Library of Congress in [26] are shown below:

- Still Image,
- Sound,
- Textual
- Moving Images,
- Web Archive,
- Datasets,
- Geospatial,
- Generic.

The first seven from this list is common with the list made by National Archives; the first four are discussed further in this section.

3.2 Domain-Oriented Metadata Standards

A number of domain-oriented metadata standards have been created. List of about 40 metadata standards was published by Digital Curation Centre [28] and list of about 35 metadata standards in Wikipedia [29]. Some universities have published similar lists, e.g. list of Domain Metadata Standards published by University of Central Florida Libraries [30]. None of those lists aspire to be complete.

Below there is given a list of metadata standards to show the variety of domains for which metadata standards have been created:

- AgMES—Agricultural Metadata Element Set (FAO); used also as namespace [31].
- AVM—Astronomy Visualization Metadata (International Virtual Observatory Alliance), standard used to tag astronomical images [32].
- PROV—W3C specification [33] designed for interchange of provenance information.
- VRA Core—a data standard for the description of images and works of art and culture (designed by Visual Resources Association 1996 [4], version 4.0 hosted by the Library of Congress [34]).
- MODS—descriptive metadata standard designed for libraries, which allows for more detailed description of resources than Dublin Core [35].
- EBUCore—EBU Core Metadata Set, which describes audio and video resources for a wide range of broadcasting applications [21].
- DIF—Directory Interchange Format—for exchanging information about scientific data sets [5].
- TEI: Text Encoding Initiative—a standard for the representation of texts in digital form, mainly in the humanities, social sciences and linguistics [36].
- ONIX—a family of metadata standards: for Books, Serials and Licensing Terms & Rights Information, designed originally to support book trade, extended to "support creating, distributing, licensing and otherwise making available intellectual property in published form, whether physical or digital" [37].
- STEP—Standard for the Exchange of Product Model Data, describing how to represent and exchange digital product information, and manage the information during life cycle of the product (officially "Industrial automation systems and integration—Product data representation and exchange" a group of standards ISO 10303) [38].

Metadata standards can be designed with a view to support long-term preservation of information. An interesting discussion concerning "Engineering-specific Metadata Requirements for Long-term Archiving" in the context of Digital Product Data Archiving and metadata standard STEP are presented in [39].

3.3 Mappings Between Metadata Standards

There is a need to make mappings between metadata standards. Several such mappings have been defined. Library of Congress and W3C have published some of them.

- MARC—Dublin Core: documents concerning the so called "crosswalk" of library standard MARC21 to Dublin Core (2001) [40] and both versions of Dublin Core— DCMES and DCMI Metadata Terms—to MARC (2008) [41].
- MODS—Dublin Core: mapping version 3 was defined in both directions on standard MODS Official Website in 2012:
  - MODS to Dublin Core—there is a comment: "In some cases, multiple MODS elements are equivalent to one Dublin Core element, and all are listed. In these cases, some metadata is lost."
  - Dublin Core to MODS—there is a comment: "In some cases, the Dublin Core element is equivalent to more than one MODS element, and all are listed. In these cases, when converting a record from Dublin Core to MODS, a decision may need to be made as to the default MODS element."
- Dublin Core (DCMES) to CIDOC CRM [42]—Conceptual Reference Model CIDOC [43] is a domain ontology in the area of cultural heritage, created by the International Council of Museums (ICOM), developed into an ISO standard 21127.
- Dublin Core to PROV mapping published by W3C [33]. PROV is a provenance metadata standard published by W3C. This mapping is made between the PROV-O OWL2 ontology and the Dublin Core Terms Vocabulary [14].

Interesting problem of mapping metadata concerns cooperation of cultural heritage institutions from various countries with Europeana [44]. Europeana uses specific format—the Europeana Data Model EDM. Europeana offers five possible routes of supplying data [45]; three of them are based on mapping. Each institution can use existing crosswalk to map and transform data directly to EDM, or to an intermediary schema or to one of standard metadata schemas, such as Dublin Core or EAD, and map to EDM using special tool (MINT). However, Europeana is oriented to sharing data and giving access on mass scale and not to long-term archiving.

3.4 Still Images

Digital still images is a basic category of assets in many digital archives. They can be used in all cases where the classic photography can be used. Digital images may play a role of copies of various paper documents, stored usually in libraries and archives, or copies of photographic images captured on films, glass plates or photographic paper. Similar status of digital copies of analogue originals may have photographies of museum artifacts. In the above mentioned cases term digitization is usually used; digital images are made by scanners and photographic cameras.

In the archives there are also stored digital photographies made for scientific, medical, commercial and other purposes. Some of those images are created by means of photographic cameras, some are results of computer analysis and transformation of measurement data.

Generally these images can be treated in similar way to ensure long-term preservation. Some differences, e.g. time of migration to new formats, may result from different formats of stored digital images.

3.4.1 Dublin Core and Other Descriptive Metadata Standards

Images can be described in Dublin Core metadata. Many digital libraries uses this standard for images. In Poland many digital libraries which are using dLibra software, describe images in DC.

Dublin Core is not well suited to describe images so some archives, libraries and museums use other metadata standards or own systems. Polish National Digital Archive (Narodowe Archiwum Cyfrowe) offers search form where user besides elements from DC have such fields as: DateFrom, DateTo, Persons visible and Persons not visible. Museums often describe their holdings in much more detailed way than DC makes possible. Some metadata can be even collection-dependent.

Librarians in previous years used to describe all their collections in MARC, even if it was difficult, e.g. in case of music records. Now they can use MODS instead [35]. Due to possibility of mapping metadata standards, e.g. MARC to MODS and MODS to DC, such institutions can offer users descriptions in two or three standards, among them is the Library of Congress.

As an example will be used descriptions of a map *New railroad map of the state of Maryland, Delaware, and the District of Columbia* available at Library of Congress [46] available in MODS and Dublin Core; MODS record is about three times as big as Dublin Core record.

Listing 1 Description of a map in Dublin Core standard (Library of Congress)

```
<dc:title>
  New railroad map of the state of Maryland,
  Delaware, and the District of Columbia. Compiled
  and drawn by Frank Arnold Gray.
</dc:title>
<dc:creator>Gray, Frank Arnold.</dc:creator>
<dc:subject>
  Railroads--Middle Atlantic States--Maps.
</dc:subject>
<dc:description>
  Shows drainage, canals, stations,
  cities and towns, counties, canals, roads completed,
  narrow gauge and proposed railroads with names
  of lines. Includes list of railroads.
</dc:description>
<dc:description>Scale 1:633,600.</dc:description>
<dc:description>LC Railroad maps, 230</dc:description>
<dc:description>
        Description derived from published bibliography.
```

```
</dc:description>
<dc:publisher>Philadelphia</dc:publisher>
<dc:date>1876</dc:date>
<dc:type>image</dc:type>
<dc:type>map</dc:type>
<dc:type>cartographic</dc:type>
<dc:identifier>
  http://hdl.loc.gov/loc.gmd/g3791p.rr002300
</dc:identifier>
<dc:language>eng</dc:language>
<dc:coverage>
  United States--Middle Atlantic States
</dc:coverage>
```

Listing 2 Fragments of description of the same map in MODS which have no equivalent information in DC

```
...
<originInfo>
  <place>
   <placeTerm type="code" authority="marccountry">pau</placeTerm>
  </place>
  ...
</originInfo>
<language>
  <languageTerm type="code" authority="iso639-2b">eng</languageTerm>
</language>
<physicalDescription>
  <form authority="marccategory">electronic resource</form>
  <form authority="marcsmd">remote</form>
  <extent>col. map 39 x 62 cm.</extent>
</physicalDescription>
...
<note type="additional physical form">
  Available also through the Library of Congress Web site
  as a raster image.
</note>
...
<recordInfo>
  <recordContentSource authority="marcorg">DLC</recordContentSource>
  <recordCreationDate encoding="marc">980520</recordCreationDate>
  <recordChangeDate encoding="iso8601">20021214203459.0</recordChangeDate>
  <recordIdentifier>5572072</recordIdentifier>
  <recordOrigin>
    Converted from MARCXML to MODS version 3.5 using
    MARC21slim2MODS3-5.xsl (Revision 1.106 2014/12/19)
  </recordOrigin>
</recordInfo>
```

Last part of additional information contained in MODS can be treated as provenance metadata.

3.4.2 Standard IPTC Photodata Core and Extension

IPTC published version 1.2 of IPTC Photo Metadata standard in 2014 [47]. However, version 1.1 of IPTC Core and IPTC Extension from 2010 is still widely used [48].

Standard was designed to be used by photographers, news agencies and publishers. It is supported by camera software of many producers and by professional photo editing software. The standard is also widely used by the amateurs.

Structure of metadata standard is simple list of terms without hierarchy and repetition. IPTC Core has four sections: *Contact, Content, Image* and *Status*:

- IPTC Contact:

  - Creator (name of the person who created the photograph),
  - Creator's Job Title (e.g. staff photographer),
  - Contact Info (Address…Phone, Email, Website);

- IPTC Image:

  - Date Created (date the photograph was created),
  - Intellectual Genre (term from *IPTC Genre Newscode*),
  - IPTC Scene Code (term from *IPTC Scene-Newscodes*),
  - Location (Sublocation, City, State/Province, Country, ISO Country Code);

- IPTC Content:

  - Headline (short synopsis of the content of photograph),
  - Description (who, what, where, when…),
  - Keywords,
  - IPTC Subject Code (IPTC Subject Newscodes taxonomy),
  - Description Writer;

- IPTC Status:

  - Title (short reference—text or numeric),
  - Job ID (can be added for transmission purposes),
  - Instructions,
  - Credit Line (how the owner should be credited when the image is published),
  - Source (original owner or copyright holder),
  - Copyright Notice (current owner or copyright holder),
  - Rights Usage Terms (how this photograph can be legally used).

IPTC Extensions defines five groups of terms: *Description, Artwork/Object, Models, Administrative* and *Rights*, which can make the description more detailed. For example the *Location* has now two fields and two meanings: *Location Created* in which the image was created and *Location Shown* in the image. IPTC Core and Extension examples are shown on page of IPTC [49]. However description of image can be limited to IPTC Core only.

In Poland this standard is used by some news agencies and commercial photographers; also by such public institutions as Museum of History of Photography [50] and History Meeting House [51].

Listing 3 IPTC metadata of exemplary photography of Museum of History of Photography (translated)

```
Caption               :
Date Created          : 1915-05-1915-08
Keywords              : MAN/PRIVATE LIFE - everyday life
  - uniform, WAR/ARMY - warfare, TYPES OF PRESENTATION
  - Portrait - group portrait
Object Name           : Two legionnaires, group portrait
```

```
Copyright Notice      : public domain, www.mhf.krakow.pl
Byline                : unknown author
```

Listing 4 IPTC metadata of exemplary photography of History Meeting House (translated)
```
Keywords              : Brandel Konrad, Gembarzewski Leszek
   - collection,  tenements, church, Warsaw,
Caption               : DI 36164; Brandel, Konrad
   (1838-1920) (photographer);  Warsaw; about 1895;
        photography; photographic paper; 11,8 x 16,3
Program               : FotoWare FotoStation
Byline                : Ligier Piotr
Caption Writer        : MM-A
Copyright Notice      : Copyright by Ligier Piotr/National Museum Warsaw
```

The IPTC Core metadata can be embedded in image files such as JPEG, TIFF and PSD either in older IIM (Information Interchange Model) data format or in newer Adobe's XMP data format. The newer one can be also embedded in DNG and PDF files.

3.4.3 Exif Standard—Technical Metadata

Exif Standard—Exchangeable Image File Format was created by the Japan Electronic Industries Development Association in 1995 (v. 1) and modified several times. Versions 2.3 and its 2.31 revision were published in 2010 and 2016 by two associations: Japan Electronics and Information Technology Industries Association and Camera & Imaging Products Association, nevertheless versions 2.2 (2002) and 2.21 (2003) are still being used in many products. A detailed information concerning revision history is enclosed in standard specification: *CIPA DC-008-Translation-2016. Exchangeable image file format for digital still cameras: Exif Version 2.31* [52].

Exif standard provided method of recording data in image files at the point of image capture; the method was defined for two file formats: JPEG (ISO/IEC 10918-1) and TIFF v. 6.0. The standard has been used in digital photo cameras and scanners, but had no provisions for information specific for scanners. Standard covers a broad spectrum of data: camera settings (such as shutter speed, focal length), date and time, thumbnail, copyright and description.

Total number of Exif metadata increased in version 2.2 to 147 [53]; in this number there were also 31 GPS tags. If camera had GPS receiver (internal or external) the GPS tags could be captured by camera software. They could also be added later by external software.

Below there are shown Exif metadata read from a photography made by SLR Sony DSLR-A700 in internal format RAW and transformed by camera software Image Data Converter SR to TIFF format and to JPEG format.

Listing 5 Exif metadata of photography made by SLR as RAW and saved in TIFF format
```
[Camera]
Camera Manufacturer       : SONY
Camera Model              : DSLR-A700
Software                  : Image Data Converter SR
Date modified             : 2015:09:18 09:06:00
```

```
[Image]
Exposure time [s]            : 1/500
F-Number                     : 5.6
Exposure program             : Aperture priority (3)
ISO speed ratings            : 200
EXIF version                 : 02.21
Date taken                   : 2014:07:22 14:15:26
Date digitized               : 2014:07:22 14:15:26
Brightness                   : 203/25
Exposure bias value          : 0
Max aperture                 : F2.8
Metering mode                : Spot (3)
Light source                 : Daylight (1)
Flash                        : No flash
Focal length [mm]            : 24
User comment                 :
FlashPix Version             : 01.00
Colour space                 : sRGB
Custom rendered              : Normal process (0)
Exposure mode                : Auto (0)
White balance                : Manual (1)
Digital zoom                 : 0
Focal length (35mm)          : 36
Scene capture type           : Standard (0)
Contrast                     : Normal (0)
Saturation                   : Normal (0)
Sharpness                    : Normal (0)
```

Exif metadata of the same photography, saved in JPEG format, contains more information.

Listing 6 Additional Exif metadata of the same photography saved by the camera in JPEG format

```
Components configuration   : YCbCr
Compressed bits per pixel  : 8
EXIF image width           : 4272
EXIF image length          : 2848
Interoperability offset    : 25460
File source                : DSC
Scene type                 : A directly photographed image
```

Standard Exif was extended in version 2.2 to cover the sound registered by photo camera. The Exif audio file specification defines method of writing audio data in files. The method was defined for RIFF WAVE Form Audio File format only. As data format is used Pulse Code Modulation (PCM) for uncompressed audio data (and also G.711, used mainly in US and Japan) and Adaptive Differential Pulse Code Modulation (IMA-ADPCM) for compressed audio data.

3.4.4 Standard MIX

Exif metadata are usually presented in form of lists similar to shown above. Originally standard MIX did not use XML notation. There have been made some efforts, e.g. Exif vocabulary workspace—RDF Schema [54] in 2003, but there was no official XML Schema.

Library of Congress published *MIX—NISO Metadata for Images in XML Schema. Technical Metadata for Digital Still Images Standard* [55]; NISO stands here for National Information Standards Organization [56]. Standard is addressed to raster

digital images. MIX Schema version 2.0 was published in 2008. Data Dictionary—
Technical Metadata for Digital Still Images (ANSI/NISO Z39.87-2006) is available
as NISO Standard.

The MIX standard contains Exif metadata from versions 2.2 and 2.21. It has
also small number of preservation metadata concerning fixity of stored image files
(group *ChangeHistory* with subgroups of metadata *ImageProcessing* and *Previ-
ousImageMetadata*. It can be applied to such file formats as JPEG2000, DjVu and
MrSID for which Exif can't be applied.

MIX standard can be used to store Exif metadata extracted from image files.
It might be useful in case of migration to new formats or could be done at the
time of ingest just for safety of embedded metadata (the last motivation is due to
the fact that popular software sometimes corrupts Exif metadata even at such simple
operations as copying). XML notation is not necessary for storing extracted metadata,
however formal structure controlled by XML schema might ensure correctness of
such operation.

3.4.5   Exif and IPTC Embedded in XMP

Format TIFF 6.0 [57] has provided space for Exif metadata and IPTC metadata (IIM).
It can also contain Extensible Metadata Platform (XMP) [58] metadata.

It should be noted that there is trend in photo camera industry to use XMP on
greater scale. XMP can be used as a container for traditional Exif and IPTC metadata. It can also be used to contain precise information concerning camera settings
technical parameters of photographs—much more precise than it was possible when
using Exif metadata.

To illustrate ability of recording metadata in XMP below are cited Exif and
IPTC metadata, and small fragments of metadata in XMP of the same photography: CopyrightInFotos-MDTest01a.jpg, published by IPTC [49].

Listing 7 Exif metadata of exemplary photography

```
[Camera]
Date modified             : 2012:04:22 20:07:21
Y Resolution              : 240
Software                  : Adobe Photoshop Lightroom 4.0 (Windows)
Image description         : Bikefestival in Wien, Rathausplatz
Camera Manufacturer       : Canon
Camera Model              : Canon EOS 60D
Copyright                 : Copyright 2012 Frank Fotofan
                            www.ffotofan.info
Artist                    : Frank Fotofan
Resolution unit           : Inch
Orientation               : top-left (1)

[Image]
Exposure bias value       : 0
EXIF version              : 02.30
Shutter speed [s]         : 1/250
Focal length [mm]         : 100
Date digitized            : 2011:04:03 13:16:34
```

```
Subject distance (m)        : 25
SubSecTimeDigitized         : 34
F-Number                    : 5.6
Focal plane Y-Resolution    : 518400/119
Focal plane X-Resolution    : 777600/181
White balance               : Auto (0)
Max aperture                : F4.2
Aperture                    : F5.6
Focal plane res. unit       : Inch (2)
Metering mode               : Multi-segment (5)
Flash                       : No flash
Exposure program            : Normal (2)
Custom rendered             : Normal process (0)
Scene capture type          : Standard (0)
```

Listing 8 IPTC metadata of exemplary photography
```
Date Created        : 20110403
Headline            : Bikefestival
Caption             : Bikefestival in Wien,
                      Rathausplatz
Copyright Notice    : Copyright 2012 Frank Fotofan
                      www.ffotofan.info
Credits             : Frank Fotofan
Location            : Rathausplatz
City                : Wien
Country             : Oesterreich / Austria
```

Listing 9 Fragment 1 of XMP metadata of exemplary photograph in RAW format
```
<rdf:Description rdf:about=""
     xmlns:aux="http://ns.adobe.com/exif/1.0/aux/">
  <aux:SerialNumber>0380227035</aux:SerialNumber>
  <aux:LensInfo>70/1 300/1 0/0 0/0</aux:LensInfo>
  <aux:Lens>EF70-300mm f/4-5.6L IS USM</aux:Lens>
  <aux:LensID>489</aux:LensID>
  <aux:LensSerialNumber>000000617b</aux:LensSerialNumber>
  <aux:ImageNumber>0</aux:ImageNumber>
  <aux:ApproximateFocusDistance>251/10</aux:ApproximateFocusDistance>
  <aux:FlashCompensation>0/1</aux:FlashCompensation>
  <aux:Firmware>1.0.5</aux:Firmware>
</rdf:Description>
```

Listing 10 Fragment 2 of rights XMP metadata of exemplary photograph in RAW format
```
<rdf:Description rdf:about=""
     xmlns:xmpRights="http://ns.adobe.com/xap/1.0/rights/">
  <xmpRights:Marked>True</xmpRights:Marked>
  <xmpRights:WebStatement>
   http://creativecommons.org/licenses/by-nc-sa/3.0/de/
  </xmpRights:WebStatement>
  <xmpRights:UsageTerms>
    <rdf:Alt>
      <rdf:li xml:lang="x-default">Creative Commons - by-nc-sa/3.0/at/</rdf:li>
    </rdf:Alt>
  </xmpRights:UsageTerms>
</rdf:Description>
```

Listing 11 Fragment 3 of XMP metadata of exemplary photograph in RAW format
```
<rdf:Description
     rdf:about="" xmlns:crs="http://ns.adobe.com/camera-raw-settings/1.0/">
  <crs:Version>7.0</crs:Version>
  <crs:ProcessVersion>5.7</crs:ProcessVersion>
```

```
<crs:WhiteBalance>As Shot</crs:WhiteBalance>
<crs:Temperature>5000</crs:Temperature>
<crs:Tint>+1</crs:Tint>
<crs:Exposure>0.00</crs:Exposure>
<crs:Shadows>5</crs:Shadows>
<crs:Brightness>+50</crs:Brightness>
<crs:Contrast>+25</crs:Contrast>
<crs:Saturation>0</crs:Saturation>
<crs:Sharpness>25</crs:Sharpness>
<crs:LuminanceSmoothing>0</crs:LuminanceSmoothing>
<crs:ColorNoiseReduction>25</crs:ColorNoiseReduction>
...
<crs:ToneCurve>
  <rdf:Seq>
     <rdf:li>0,  0</rdf:li>
     <rdf:li>32,  22</rdf:li>
     <rdf:li>64,  56</rdf:li>
     <rdf:li>128,  128</rdf:li>
     <rdf:li>192,  196</rdf:li>
     <rdf:li>255,  255</rdf:li>
  </rdf:Seq>
</crs:ToneCurve>
<crs:ToneCurveRed>
  ...
</crs:ToneCurveRed>
...
</rdf:Description>
```

The examples presented above show great ability of XMP (using also RDF) to record with great precision parameters of photographs and camera settings, beyond the scope of Exif standard.

However, this ability does not improve anything with respect to long-term archiving. Format RAW—as commercial—is not accepted in this role, the more so there is no one RAW format but many different formats of various producers.

Digital Negative (DNG) format can be used in the archives. It is an open lossless format, written by Adobe [59], designed for archiving RAW photographs taken by cameras of various manufacturers. Photographs saved in different versions of RAW format can be converted to one DNG format. Using this DNG format and converters is royalty free. DNG format contains checksum information important in archiving. In the Format Guidance for the Transfer of Permanent Electronic Records [27] the DNG format (specification version 1.4.0.0) has status of acceptable format for category *Digital Photography*.

3.5 Sound Archives

3.5.1 Introduction

Sound archive or audio archive can be a part of collections of a library, an archive or museum, e.g. British Library Sound Archive. Sound archive can also be a separate institution, e.g. radio archive. Sound archives from many countries created in 1969 *International Association of Sound Archives (IASA)* [60], which in 2012 extended range of collections, changing name to *International Association of Sound and Audiovisual Archives* but leaving the the abbreviation IASA unchanged.

IASA published i.a. the *IASA Cataloguing Rules (1999)* [61] and the *Guidelines on the Production and Preservation of Digital Audio Objects* (2009) [62].

3.5.2 IASA Cataloging Rules

The IASA Cataloging Rules define 11 groups of metadata elements:

0:   Preliminary notes,
1:   Title and statement of responsibility,
2:   Edition, issue, etc.,
3:   Publication, production, distribution, broadcast, etc., and date(s) of creation,
4:   Copyright,
5:   Physical description,
6:   Series,
7:   Notes,
8:   Numbers and terms of availability,
9:   Analytic and multilevel,
10: Item/copy information.

Physical description concerns physical media, such as tapes, records and disks; large part of archives' resources were using such media in 1999.

Each of 11 groups contains several elements, e.g. group no 3 *Publication, production, distribution, broadcast, etc., and date(s) of creation* contains:

3.0.   Scope and definitions,
3.A.   Preliminary rule,
3.B.   General rule,
3.C.   Place of publication, production, distribution, broadcast, etc.,
3.D.   Name of publisher, producer (production company), distributor, broadcaster, etc.,
3.E.   Optional addition. Statement of function of publisher, producer (production company), distributor, broadcaster, etc.,
3.F.   Date of publication, production, distribution, broadcast, etc.,
3.G.   Place, name and date of manufacture,
3.H.   Date(s) of creation (unpublished items only),
3.I.   Reproductions (which are themselves unpublished).

Each of above elements has brief description and explanation.

From 37 examples given by IASA, with various types of content, like popular music, classical music, opera, oral history, spoken word, radio music etc. on various types of media, the following two have been chosen:

Example 1. Radio music production on DAT—live recording

Johannes Brahms und seine Freunde (1)/Markus Brändle, producer; Erich Heigold, sound engineer; Gertrud Bastuck, cut.—Saarbrücken: SR2, 26.12.1988, (13.05–4.30).—Copyright: SR; GEMA; GVL.—1 sound tape (DAT, 84 min 46 s): digital (AAD), stereo, AES/EBU standard.—(Der musikalische Salon) Rolf Sudbrack, author and speaker; Jewgenij Koroliov, Ljupka Hadzigeorgiev, piano. Recorded live 27.01.1988, Saarbrücken (Germany), Funkhaus Halberg (Grosser Sendesaal) Sampling frequency and quantisation: 48 kHz, 16 bit. Contents: Variationen über ein Thema von Robert Schumann, Es-dur. op.23 (Leise und innig) / Johannes Brahms (1 min 49 s)—4 Balladen op.10 (Nr.3 h-moll; Nr.4 H-dur)/Johannes Brahms (4 min 30 s.; 9 min 36 s.)—3 Romanzen op.21 (Nr.1 a-moll; Nr.2 F-dur)/Clara Schumann (5 min 28 s; 1 min)—4 Klavierstücke op.2 (Nr.2 Kanon)/Albert Dietrich (4 min 20 s)—Sonate für Klavier zu 4 Händen g-moll, op.17 (2.Satz)/Hermann Goetz (5 min 33 s)—Intermezzo B-dur, op.76 Nr.4/Johannes Brahms (2 min 29 s)—Capriccio d-moll, op.116 Nr.7/Johannes Brahms (2 min 10 s). Copy from 2 tape reels, analogue: stereo; 38 cm/s, Telcom C4.

Example 2. Opera film

Aida [videorecording] / music: Giuseppe Verdi; original libretto: Antonio Ghislanzoni; produced by: Staffan Rydn; directed by: Claes Fellbom; revised text and screenplay: Claes Fellbom; costume designer: Inger Pehrsson; art director: Lotta Melanton; choreography: Ann-Charlotte Lindström.—Sweden: the Swedish Film Institute [distributor]: Isis Film, the Swedish Film Institute, Sveriges television TV2 and Filmhuset [production companies], [1993].—1 videocassette (VHS, ca. 122 min): sd. (stereo), col. Credits: Berndt Fritiof/Filmmixarna, sound/mix supervisor; Hans Ewers, sound/music supervisor; Jörgen Persson, director of photography; et al. Cast: Margareta Ridderstedt (Aida), Niklas Ek (Radames), Robert Grundin (Radames' voice), Ingrid Tobiasson (Amneris), Jan van der Schaaf (Amonasro), Alf Häggstam (Ramfis), Staffan Rydén (Pharao's spokesman), et al. The Swedish Folkopera Orchestra and Choir; Kerstin Nerbe, conductor. Filmed version of Verdis Aida, shot in Lanzarote, Spain. This production was originally staged by the Swedish Folkopera.

In the above examples descriptive metadata are used together with technical metadata and sometimes also provenance metadata. Relations between persons or institutions and their roles, e.g.

Jewgenij Koroliov, Ljupka Hadzigeorgiev—piano;
Giuseppe Verdi—music;
the Swedish Film Institute—distributor;
Jörgen Persson—director of photography (credits);
Niklas Ek—Radames (cast);

represent significant part of information concerning the archives' holdings.

3.5.3 Acceptable Formats for Audio

The Format Guidance for the Transfer of Permanent Electronic Records [27] contains following recommendations concerning preferred or acceptable formats for Digital Audio:

- Preferred formats:

  - Broadcast Wave (BWF); codec LPCM (Linear Pulse Code Modulated Audio); version 1 and version 2; EBU Tech specification 3285,
  - Free Lossless Audio Codec (FLAC); version 1.21.

• Acceptable formats:

  – Audio Interchange Format (AIFF); codec LPCM (Linear Pulse Code Modulated Audio); v. 1.3,
  – MPEG Audio Layer III (MP3); codec MP3enc; ISO/IEC-11172-3 part 3—Audio [63],
  – Waveform Audio File Format (Wave); codec LPCM (Linear Pulse Code Modulated Audio).

General technical recommendations: sample rate at least 44.1 kHz, but 96 kHz is encouraged; minimum of 16 bits per sample, but 24 bits per sample is encouraged.

3.5.4 Audio Metadata Standards

New edition of *Guidelines on the Production and Preservation of Digital Audio Objects* (2009) [62] is based on the Reference Model for an Open Archival Information System (OAIS). Discussion and recommendations concerning metadata take into account aspects of preservation, e.g. "the record of the creation of the digital audio file, and any changes to its content, must be created at the time the event occurs. This history metadata tracks the integrity of the audio item and, if using the BWF format, can be recorded as part of the file as coding history in the BEXT chunk. This information is a vital part of the PREMIS preservation metadata recommendations." In these guidelines there are mentioned such metadata standards as Simple Dublin Core, DCMI Metadata Terms with Applications Profiles, AES57 (Audio Engineering Society), METS and PREMIS. Full name of AES57 is: *AES57-2011: AES standard for audio metadata—Audio object structures for preservation and restoration*. Standard provides vocabulary.

Relations between persons and their roles can hardly be expressed in Dublin Core. It is suggested in the Guidelines to use MARC Code List for Relators [15] and DCMI Terms, as in the example below:

```
<dcterms:contributor>
  <marcrel:CMP>Giuseppe Verdi</marcrel:CMP>
  <marcrel:CNG>J\"orgen Persson</marcrel:CNG>
</dcterms:contributor>
```

Verdi is tagged here as composer (CMP), Persson as cinematographer (CNG), what is equivalent to director of photography. However, not all roles mentioned in the above examples are on the relators list, e.g. there is no piano (pianist etc.), only general term musician. Expressing casting in Dublin Core is even more difficult. IASA Guidelines suggest that in such cases MODS is a better standard than DC.

In this context EBU Core Metadata Standard [21] can also be taken into account. In the first version this standard [9] (2001) EBU decided to add refinements to elements from simple Dublin Core list. One of the refinements was *Role* which was added to elements: *Creator* and *Contributor*. The EBU Reference Data Tables: Roles in broadcasting was also added to these elements as recommended Encoding scheme for *Role*, with the following *EBU Comment*: "The element refinement Role is added by EBU and not part of standard DCMES. The content of the qualifier Role must be taken from a controlled list of authorized roles. It is recommended that roles are taken from the EBU Reference Data Table [6], but this list can be extended to cover special local needs."

EBU Comments were added to all elements. Element *Subject* got a comment: "Persons as subjects are also placed here. Genre of the content is placed under element Type"; a few comments concern "Recommended best practice". A few *Element Encoding schemes* were added, among them *EBU Reference Data Table: Type of resource* and *RDS: PTY display terms* for element *Type* and *SMPTE Unique Material Identifier (UMID)*, *International Standard Recording Code (ISRC)* and *International Standard Audiovisual Number (ISAN)*.

New version of EBU Core [21] (2015) recommends using EBU Classification Schemes and ontologies [64]. Roles are defined in *EBU Role Code Classification Scheme* (2013) [65]. List contains about 680 terms—much more than list of *MARC Code List for Relators.*

3.5.5 Formats and Embedded Metadata

Basic recommended format for archiving the sound is WAVE (extension WAV). It is designed to store audio bitstream in chunks (a specific container format). Most often WAV files contain uncompressed sound in the Linear Pulse Code Modulation (LPCM) format, however are also able to contain compressed sound. WAV can contain metadata (tags) in the INFO chunk; it can embed any kind of metadata.

There is also version of WAV called BWF (Broadcast Wave Format), specified by EBU in 1997 as EBU Tech 3285, updated in 2001 (when SMPTE UMID was added) and revised in 2011. BWF has extension chunk BEXT designed for broadcasters needs. Format is specified for use in AES31 and also by SMPTE 382M (Mapping of AES and Broadcast Wave audio into the MXF generic container). In 2012 BWF became able to contain identifiers, especially the International Standard Recording Code ISRC (EBU Tech 3352: The Carriage of Identifiers in the Broadcast Wave Format BWF).

The MP3 specification is part of the MPEG-1 standard (1992) designed for lossy compression of video and audio. MPEG-1 standard is published as ISO/IEC 11172—*Information technology—Coding of moving pictures and associated audio for digital storage media at up to about 1.5 Mbit/s.* Part III of the standard, which concerns Audio, specifies three layers: layer I,layer II and Layer III. MP3 is an abbreviation of *MPEG-1 Layer III.* MP3 is a lossy compression of audio and as such is not recommended to use for long-term archiving. However it became extremely popular and it may happen that some MP3 files will be sent to long-term archives (if there is no uncompressed version). MP3 specification does not define tag formats, however there are *de facto* metadata standards. Metadata are usually embedded at the end or at the beginning of MP3 file.

In version 1 of the standard last 128 bytes of file contained 6 fields, which might be converted to Dublin Core:

- Song Title—30 bytes—Title,

- Artist—30 bytes—Contributor,
- Album—30 bytes—Source,
- Year—4 bytes—Date,
- Comment—30 bytes—Description,
- Genre—1 byte—Subject (to be decoded).

In 1998 was created Version 2 (ID3v2) of the standard, which introduced radically different structure of metadata. Metadata information is added at the start of the file, not at the end. Fields (tags) are of variable size, not fixed. Each field has own identifier (four chars in versions 2.3 and 2.4). Many fields were added; total number increased to 74 in version 2.3 and 83 in version 2.4. In most of these fields there is descriptive and technical metadata. There are also fields for commercial information and for user defined text information.

Information from basic descriptive tags can be easily mapped to Dublin Core, e.g.

- MCDI—Music CD Identifier—to Identifier;
- TALB—Album/Movie/Show title—to Source;
- TCOM—Composer—to Contributor (Role);
- TCOP—Copyright message—to Rights;
- TDRC—Recording time—to Format;
- TEXT—Lyricist/Text writer—to Contributor (Role);
- TIT2—Actual name of the piece (e.g. "Adagio")—to Title;
- TLAN—Language—to Language;
- TPE1—Lead artist/Lead performer/Soloist—to Contributor;
- TPE3—Conductor—to Contributor (Role);
- TPUB—Publisher—to Publisher;
- TSRC—ISRC (International Standard Recording Code)—to Identifier;
- WCOP—Copyright/Legal information—to Rights.

Standard ID3 was designed to be used in MP3, but can also be used in other formats, i.a. WAV and MP4.

3.6 Cinema and TV Archives

3.6.1 Introduction

FIAF—Fédération Internationale des Archives du Film/International Federation of Film Archives [66] was founded in 1938 in Paris by four film archives from US, UK, France and Germany. In 2015 FIAF had 155 affiliates (85 members and 70 associates) in 74 countries. TV archives were created later, mainly by TV broadcasters. European Broadcasting Union had no special forum for TV archive issues. A few archives which recognized their need of collaboration and sharing expertise created in 1976 IFTA—International Federation of Television Archives [67].

The need of various forms of cooperation among audiovisual archives motivated creation of the *Co-ordinating Council of Audiovisual Archives Associations* (CCAAA) which was founded at the end of the 1990s. Besides FIAF and IFTA the participants of CCAAA were the International Council of Archives (ICA), the International Association of Sound Archives (IASA), and the International Federation of Library Associations (IFLA).

Among fields of common interest were cataloging and (later) metadata standards, which were earlier developed within libraries and 'traditional' archives.

3.6.2 FIAF Cataloging Rules

FIAF published Cataloging Rules for Film Archives in 1991 [68]. These rules were based on libraries experience in bibliographic description: International Standard Bibliographic Description for Non-Book Materials (ISBD (NBM)). The approach to describing films predates the era of digital production. It is focused on 'reels' of film and their localization in the archive. Below there is exemplary description shown in [68].

Listing 12 Exemplary description of film based on 1991 rules (fragments)

```
TITLE         The Princess' Necklace
STATEMENTS OF RESPONSIBILITY
              director, Floyd France ;
              story, Clare Freeman Alger ;
              scenarist, E. Clement D'Art
PRODUCTION    US : Thomas A. Edison, Inc.
DISTRIBUTION  [producer], 1917 ;
ETC.          US : K.E.S.E. [distributor], 1917
COPYRIGHT     (c): US: Thomas A. Edison, Inc. 31 Aug 17;
              LP11335.
              Viewing print: 4 reels of 4(1498 ft.):
              16 mm.: S., b&w, si. / USW FLA 1742-1745.
              Duplicate negative: 4 reels of 4(1498 ft.);
PHYSICAL      16mm. : S., b&w, si. / USW FRA 4336-4339.
DESCRIPTION   Archival positive: 4 reels of 4(1498 ft.);
              16mm. : S., b&w, si. / USW FRA 4340-4343.
SERIES        (CONQUEST PROGRAM ; NO. 8)
              Cast: William Calhoun, Kathleen Townsend,
              Wallace MacDonald, Susan Mitchell,
              Dorothy Graham, Roy Adams.
NOTES         Summary: A fairy tale in which a stranger
              comes to Happyland in order to learn the
              master secret of happiness (...)
              References: Moving Picture World, (...)
```

New approach to cataloging films, based on metadata, was published by FIAF scarcely in 2016 in *The FIAF Moving Image Cataloging Manual* [69]. Film archives couldn't wait from 1991 to 2016. They had to implement other standards. Authors of this document made comments concerning relationship of new document to standards which were applied in the meantime. One comment refers to European Standards Committee standards EN 15744 and 15907. s comment concerns i.a. the EBU standard: "While these guidelines are intended to be applicable to all forms of moving image materials, archives with extensive broadcasting collections may wish to look to broadcasting-specific metadata schemas such as EBUCore and PBCore for additional guidance."

3.6.3   CEN European Standards Committee

European Standards Committee (CEN) published two Cinematographic Works Standards: EN 15744 and EN 15907 which define the metadata essential for facilitating data exchange and consistent identification of moving images [70]. Standards are available at [71, 72].

European metadata standard EN 15744 defines following minimum set of metadata for cinema films:

- Title,
- Series/Serial,
- Cast,
- Credits,
- Production Company,
- Country of Reference,
- Original Format,
- Original Length,
- Original Duration,
- Original Language,
- Year of Reference,
- Identifier,
- Genre,
- Relationship,
- Source.

Simplicity of this list reminds that of Dublin Core, but there are differences. The meaning of each data element is explained in [73]. Three elements relate to physical parameters of "the first known manifestation of a cinematographic work" (original format, length, duration).

Metadata standard 15907 [74] has more complex data model which allows to make much more detailed descriptions.

- Primary Entities (Cinematographic Work, Variant, Manifestation, Item, Content).
- Contextual Entities (Agent, Event).
- Elements (Identifier, Record Source, Title, Identifying Title, Country of Reference, Year of Reference, Format, Extent, Language, Production Event, Publication Event, Award, Decision Event, IPR Registration, Preservation Event, Subject Terms, Content Description).
- Common Element Types (Region, Timespan, Language Tag).
- Relationships (HasAgent, HasEvent, HasContent, HasAsSubject, HasOtherRelation, HasVariant, HasManifestation, HasItem).

3.6.4   Comment on FIAT/IFTA Cataloging Rules

FIAT/IFTA published in 1992 a minimum list of 22 fields for cataloging broadcast materials [75]. The list contained three groups of fields: descriptive fields (i.a. identification, personnel information, keywords, time and place of shooting), technical (physical description, date of transmission) and legal. List was available for members of FIAT/IFTA only.

3.6.5   Metadata in TV Archives—EBU Survey

Standard EBUCore is used by a number of TV archives. The EBUCore was discussed in Sect. 3.5.4 as standard concerning audio assets. However this standard concerns both audio (radio) and TV assets. It can be easily noticed e.g. in *EBU Role Code Classification Scheme* (2013) [65], where list of codes was created by joining the MPEG7 cast roles plus TV-Anytime Role Classification System [76] (TV-Anytime is also managed by EBU).

EBUCore was also indicated in the new FIAF Moving Image Cataloguing Manual [69]. It could be done because FIAF extended scope of Manual from films to TV. Together with EBUCore was indicated standard PBCore—Public Broadcasting Metadata Dictionary, developed in US to describe sound and moving images, digital and analog. Authors claim that: "Because it is so useful in describing media assets, a growing number of film archives and media organizations outside of public broadcasting have adopted PBCore to manage their audiovisual assets and collections [77]." Like EBUCore, the PBCore is based on Dublin Core. Version 2.1 of PBCore was released in 2015.

Another standard should be mentioned here: P/META standard [78]—managed by EBU— "originally designed to support business to business content exchanges, it has also been implemented for other purposes like for exchange between production systems or as high level descriptive semantic metadata".

A number of metadata standards for TV is registered in SPTME Metadata Registry class 13–14 [79] managed by SMPTE Registration Authority, LLC. Class 13 is reserved for metadata registered by an organization for public use. Among 14 organizations which registered metadata in this class are: EBU, Pro-MPEG Forum, Association of Moving Image Archivists, Public Broadcasting Service (PBS), Audio Engineering Society (AES) and Library of Congress.

EBU prepared an *Archive Report 2010* [80] based on a survey concerning TV archive-specific issues. One of them concerned the metadata standard.

Archives were divided into two groups:

- archives with integrated file-based facilities—advanced,
- archives with partially file-based facilities—beginners.

Two main groups of users were Broadcasters and Vendors. Two groups of metadata standards were standards for exchange and standards for internal use. Main results of the survey:

- Metadata standards for exchange:

  - In the group of archives with *integrated file-based facilities* "almost 20% of Broadcasters used Dublin Core based standards for exchange between archive and production" and 13% used their own internal metadata scheme. The majority of Vendors used schemes according to user requirements.
  - "Broadcasters with *partially file-based facilities* either specify their own in-house formats (most popular) or use a Dublin Core based or P/META based Metadata scheme for exchange."

- Metadata standards for internal use:

  - "Broadcasters with *integrated file-based facilities* indicate mainly the use of their own proprietary formats (most popular—25%) or use Dublin Core based formats (13%)"; small number of archives use P/META based Metadata. The majority of Vendors used schemes according to user requirements.
  - "Broadcasters with *partially file-based facilities* report similar results" as for exchange.

Authors of the survey conclude: "In-house formats are more frequently specified for internal archive usage, whereas Dublin Core based formats serve mainly for Metadata exchange between archive and production."

3.6.6 Comments on File Formats

According to [27] there is given one preferred format for Digital Cinema, i.e. Digital Moving Picture Exchange Bitmap (DPX), specified by SMPTE 268M; accepted formats are not specified in this case.

As to Digital Video, preferred formats are not specified in [27]; there are indicated following accepted formats:

- Audio Video Interleaved Format (AVI)—uncompressed 4:2:2,
- QuickTime File Format (MOV)—uncompressed 4:2:2,
- Windows Media Video 9 File Format (WMV)—codec: VC-1,
- MPEG 4—codec: H.264,
- MPEG-2 Video (MPEG2),
- Material Exchange Format (MXF)—J2K-losslessly-compressed.

In Poland Apple ProRes 422 HQ is also recommended; MXF is not recommended (not popular).

4 Rights Metadata

Many metadata standards are able to contain information concerning intellectual rights to digital objects. For example Dublin Core has a field *Rights* in which simple information without qualifiers can be stored. More complex rights information can be stored in IPTC, but this standard is rather unsuitable to be used outside the commercial photography.

Rights metadata are needed in many domains. Various standards of rights metadata have been created. Among worth mentioning is Open Digital Rights Language (ODRL) [81], being developed under auspices of W3C (in 2015 version 2.1 of Core Model, XML Encoding, Common Vocabulary and Ontology). As an example of commercial standards can be indicated group of standards created by consortium DDEX [82]. Consortium defined group of over 20 standards DDEX used to support exchange of information among the media companies, music licensing organizations, digital service providers etc. However, the long-term preservation does not need such detailed information concerning rights.

Among standards deserving more attention in context of long-term archiving are:

- METS, which can be used to create packages SIP, and is designed to contain rights metadata;
- PREMIS, which defines characteristics of rights concerned with preservation activities and gathers information during storage of digital objects in the archive.

Rights metadata in standard METS are defined in external XML Schema: METSRights.xsd [83]. There are three main elements defined:

- RightsDeclaration—description of intellectual property rights associated with digital object or its part,
- RightsHolder—a person or organization which owns some intellectual property rights,
- Context—a description concerning who has what permissions and constraints.

In the file METSRights.xsd are also given comments concerning metadata and examples. Below is one of examples:

Listing 13 Exemplary metadata in standard METSRights (fragments)
```
<RightsDeclaration>
  Any re-use of these materials in publication may
  only be done with the explicit permission of the
  Charles L. Dodgson Estate. Please contact the
  Fales Library staff if you wish to use any of
  these materials.
</RightsDeclaration>
...
<RightsHolder RIGHTSHOLDERID="FALESRH01">
  <RightsHolderName>The Estate of Charles L.
    Dodgson (Lewis = Carroll)</RightsHolderName>
  <RightsHolderComments>
   The estate of Charles Dodgson is represented by
   AP Watt agency of London. All permissions issues must
   be addressed to them.
  </RightsHolderComments>
  <RightsHolderContact>
    ...
  </RightsHolderContact>
</RightsHolder>
...
<Context CONTEXTCLASS="GENERAL PUBLIC">
  <Permissions OTHER="false" PRINT="false"
    DELETE="false" MODIFY="false"
```

```
    DUPLICATE="false" COPY="false"
    DISPLAY="true" DISCOVER="true"/>
  <Constraints CONSTRAINTTYPE="QUALITY">
    <ConstraintDescription>
      Users may only access digital copies of photographic
      materials digitized at 50 dpi or less.
    </ConstraintDescription>
  </Constraints>
</Context> <Context CONTEXTCLASS="REPOSITORY MGR">
  <Permissions OTHER="false" PRINT="true"
    DELETE="true" MODIFY="true"
    DUPLICATE="true" COPY="true"
    DISPLAY="true" DISCOVER="true"/>
</Context>
```

The above informations concern the current situation. There is some lack of information concerning problems which might appear in the future with respect to preservation activities.

Standard PREMIS is based on entity semantic units: Object, Event, Agent and Rights.

Data Dictionary version 2.2 describes the *Rights Entity* using following terms:

Listing 14 Rights metadata for Entity Semantic Units in standard PREMIS (Data Dictionary version 2.2)—chosen terms

```
rightsStatement
rightsStatementIdentifier
rightsBasis
copyrightStatus
copyrightJurisdiction
copyrightDocumentationIdentifier
copyrightApplicableDates
startDate
endDate
licenseDocumentationIdentifier
licenseDocumentationRole
licenseTerms
licenseApplicableDates
statuteInformation
statuteJurisdiction
statuteCitation
statuteDocumentationIdentifier
statuteApplicableDates
otherRightsInformation
rightsGranted
act
restriction
termOfGrant
termOfRestriction
linkingObjectIdentifier
linkingObjectRole
linkingAgentIdentifier
linkingAgentRole
rightsExtension
```

This short list gives an idea what could be expressed by such terms.

Choice between PREMIS and METS in case of rights metadata should be made carefully. Library of Congress published Guidelines for using PREMIS with METS [84], which may be helpful.

5 Delivery of Content Data and Metadata to Archive

5.1 Packages and Metadata Standards

The OAIS reference model defines three types of information packages:

- Submission Information Package (SIP)—information delivered from the content provider to the archive,
- Archival Information Package (AIP)—information stored by the archive,
- Dissemination Information Package (DIP)—information delivered to a user on request.

Generally packages contain:

- digital objects, i.e. content files and metadata files (or embedded metadata) concerning objects or individual files, and
- metadata concerning the package itself.

Metadata concerning the package should inform at least about the structure of the package: e.g. list of files, formats of files, structure of catalogs etc. and about such attributes of files as checksums. Some such system of metadata became standards. It usually means that they have been defined in detailed and precise way and have syntax controlled by XML Schema which is maintained by an organization responsible for the standard.

Several such 'packaging' metadata standards have been developed, among them:

- METS: Metadata Encoding and Transmission Standard, designed for libraries (version 1.1 in 2002) [6],
- XFDU: XML Formatted Data Unit, designed for space data (ISO standard 13527 in 2003) [85],
- LOTAR: Metadata for Archival Package Workgroup [86],
- E-ARK: European Archival Records and Knowledge Preservation [87].

METS and XFDU achieved greater significance than other two and are presented in this chapter in more details.

METS was used in many projects. METS Implementation Registry [88] contains list of about 40 projects from 8 countries. List of METS registered profiles [89] extends this list further.

Information concerning the use of XFDU is gathered in different way [90]. CCSDS gathers information about missions which are "known to be using CCSDS-recommended protocols". Showing list of over 800 such missions CCSDS claims that "many of these missions also follow CCSDS Recommendations for data archiving" what can mean that they use XFDU. Among projects which are known for using XFDU two European projects can be mentioned:

- project CASPAR [91]—Cultural, Artistic and Scientific knowledge for Preservation, Access and Retrieval—an EU Integrated Project,

- project SAFE—Standard Archive Format for Europe [92].

Requirements concerning packages are not limited to metadata standard. They become more specific for individual projects. The project KOPAL can be used as an example. The German National Library in cooperation with other German library developed an archiving and exchange package format for digital objects—called Universal Object Format (UOF). UOF package consists of content files packed to one file (ZIP or TAR) and a metadata file created according to METS Schema. Among obligatory attributes of files are:

- ID,
- MIMEtype,
- Created,
- Size,
- Checksum,
- ChecksumType.

Detailed requirements are formulated in *Co-operative Development of a Long-Term Digital Information Archive, Frankfurt am Main* 2006 [93].

The packages are transformed in the archive. During ingest process SIP is transformed to AIP or to a part of AIP, or to several AIPs. AIP can be enriched by information concerning preservation activities undertaken by the archive. Archival information requested by a user is prepared by transformation of appropriate AIP or AIPS to DIP.

Archival packages can also be used to transfer information between repositories. DIP sent from one archive can be accepted in another as SIP. Sending AIP directly to another archive is also considered. In both cases some details concerning construction of packages must be agreed between archives. A detailed discussion of such problems concerning transfer of packages is presented in *Enriched Archival Information Packages* [94].

Project LOTAR—Metadata for Archival Package Workgroup [86] claims that their objective is "is to develop, publish and maintain a standard for Metadata for Archive Packages in a neutral form that can be read and reused independently of changes in the IT application environment originally used for creation".

E-ARK project—European Archival Records and Knowledge Preservation—is oriented towards big data. Authors claim: "Special attention is paid to the requirement of being able to handle very large data sets which have to be split into several parts because it is not possible to archive them as a single coherent unit [87]".

5.2 Metadata Encoding and Transmission Standard (METS)

METS standard is maintained by the Library of Congress [6]. It was created to be used mainly by libraries. It is an XML Schema which allows for:

- expressing the structure of a digital entity by identifying files which comprise the content of the entity,
- linking descriptive metadata and other metadata with digital content,
- wrapping digital content and associated metadata.

Document METS [95] have following sections:

1. METS header,
2. descriptive metadata,
3. administrative metadata,
4. file section,
5. structural map section,
6. structural link section,
7. behavior section.

The header contains information concerning the METS document itself (identifiers, dates, identifiers of other sections or elements for internal linking, agents).

Descriptive metadata can refer to the METS object as a whole or to any of its components. METS does not define own descriptive metadata. Many descriptive metadata standards can be used, e.g. Dublin Core, MARC, MODS, VRA, EAD; local standards can also be used. Multiple sections are allowed; they can use different standards. METS provides a means to link metadata to digital content or to other metadata, e.g. administrative or structural (<mdRef>). Links can be defined also to non descriptive metadata, e.g. to preservation metadata PREMIS. There is also wrapper element defined (<mdWrap>) which can contain either Base64 encoded metadata or XML encoded metadata. Metadata Reference element mdRef can point to external location.

Administrative section contains metadata which refer to the object as a whole, to its components or to original source material. The section is divided into four subsections that contain:

- technical metadata,
- intellectual property rights,
- analog/digital source metadata,
- provenance metadata.

Way to handle these metadata is similar to that of descriptive type. METS does not define own standard of these metadata types. Linking and referencing can be made in the same way.

File section acts as inventory of content files that comprise the digital object described in the document. There is a <file> element for each file; a <fileGrp> element allows files to be grouped. Element <file> has attributes such as:

- ID—identifier unique within the METS document,
- MIMETYPE,
- SEQ—sequence of this file relative to the others in its group <fileGrp>,
- SIZE,

- CREATED,
- CHECKSUM—the checksum value,
- CHECKSUMTYPE—the checksum type (it must be one from the list accepted in the current version of METS Schema),
- OWNERID,
- ADMID—contains the ID attribute value of the `<techMD>`, `<sourceMD>`, `<rightsMD>` or `<digiprovMD>` elements within the administration section pertaining to the file,
- DMDID—contains the ID attribute value of the `<dmdSec>` pertaining to the file,
- GROUPID—relation to other file groups,
- USE.

Structural map organizes files in a consistent structure, e.g. hierarchical. There can be more than one structural map, e.g. one can determine logical structure of a book (chapters, sections, etc.) and the other physical structure (sequence of pages). In general structural map can link files containing text, sound, still images, video and others. Structure is described as a tree of nested elements `<div>` (division); links by pointers `<fptr>` and `<mptr>`.

Structural map is obligatory in METS document.

A number of Example Documents is presented on METS page [96]. The examples concern various types of digital objects or different domains, e.g. Bibliographic Record, Image with Text, Image with Video, Video with Transcript, Sheet Music, Page Turners, Maps & Geographic. Size of METS metadata files varies from about 10 kB to 200 kB, exceeding 1 MB in one case.

5.3 XML Formatted Data Unit (XFDU)

XFDU standard was created by *Consultative Committee for Space Data Systems* (CCSDS), international organization founded by space agencies for the development of communication and data handling standards. Member agencies come from such countries as USA, Russia, China, Japan, United Kingdom, France, Germany, Italy, Canada and Brazil. Also European Space Agency ESA is the member of CCSDS. About 30 other agencies from various countries have the observer status.

Main results of CCSDS activities are published as *Recommendations*, e.g. Recommended Standards or Recommended Practices. Recommended standards are based on consensus of the Committee members. In case of XFDU packages three standards have special meaning:

1. *XML Formatted Data Unit (XFDU). Structure and Construction Rules. Recommended standard. CCSDS 661.0-B-1. Blue Book, September 2008. ISO equivalent 13527:2010* citeGP:xfdu;
2. *Producer-Archive Interface Methodology Abstract Standard (PAIMAS). Recommendation for Space System Practices. CCSDS 651.0-M-1. Magenta Book. May 2004. ISO equivalent 20652:2006* [97];

3. *Producer-Archive Interface Specification (PAIS). Recommended Standard. Blue Book. Issue 1. February 2014. CCSDS 651.1-B-1. ISO equivalent 20104:2015* [98].

As it is defined in those standards, standard 1 (XFDU) "is a technical Recommendation to use for the packaging of data and metadata, including software, into a single package (e.g., file, document or message) to facilitate information transfer and archiving." The purpose of Recommendation 2 "is to identify, define and provide structure to the relationships and interactions between an information Producer and an Archive". Recommended Standard 3 "is a technical Recommendation providing the abstract syntax and an XML implementation of descriptions of data to be sent to an archive." It includes "one concrete implementation for the packages based on the XML Formatted Data Unit (XFDU) standard".

XFDU packaging is based on three concepts:

- Package Interchange File,
- Manifest Document,
- XFDU (XML Formatted Data Unit).

Package Interchange File is a container, in which the special file called Manifest Document must be included and can be bundled a number of other files. In the Manifest Document relations among the files are described and locations of all the files within the Package Interchange File are indexed. The Manifest can also contain files (data and metadata files) and have references to external files. Term XFDU is referred to Package Interchange File together with external files and packages referenced from the Manifest file.

XFDU functionality contains a few features: two packaging techniques are indicated in the XFDU standard: single XML document and archive file, and the XML messaging form is provided in the future, e.g. XML-binary Optimized Packaging XOP [99].

There are three Metadata/data linkage options:

- inclusion in Manifest as base64 or XML,
- referenced directly as binary or XML,
- referenced or included as Data Object.

There is one assumption concerning the encoding and transformations: "The ability to allow/reverse multiple transformations on files".

The XFDU Manifest file can include:

- Package Header—contains metadata which apply to the whole package;
- Information Package Map—uses ContentUnit elements which contain pointers to data objects and to their metadata;
- Data Objects—a byte stream and data necessary to reverse the transformation and return to original format;
- Metadata Objects—metadata can be recorded separately for each item;
- Behavior Objects (mainly for future use).

Package designer can define own metadata model. There are also predefined metadata which support the OAIS information model.

The construction and use of XFDU standard is illustrated by a number of examples [85].

6 Preservation Metadata

6.1 PREMIS Standard

PREMIS achieved great significance among preservation metadata standards. The name PREMIS stands for *PREservation Metadata: Implementation Strategies*. Originally it was name of international working group established in 2003 to design metadata for use in digital preservation. This group released in 2005 report: *Data Dictionary for Preservation Metadata: Final Report of the PREMIS Working Group (Version 1.0)* [100]. Project Credo used version 2.2 of PREMIS Data Dictionary, released in 2012 (next version 3.0 was not complete at that time). All three versions of standard PREMIS are available at [7]. Description of standard PREMIS in this chapter is based on version 2.2 of Data Dictionary for Preservation Metadata [101].

Construction of metadata in PREMIS standard is based on *data model* and *semantic units*.

Data model defines five interrelated (*Entities*): *Intellectual Entity, Object, Event, Agent* and *Rights*. The Intellectual Entity is defined as "a set of content that is considered a single intellectual unit for purposes of management and description". Other entities have following definitions:

- Object (or Digital Object): a discrete unit of information in digital form.
- Event: an action that involves or impacts at least one Object or Agent associated with or known by the preservation repository.
- Agent: person, organization, or software program/system associated with Events in the life of an Object, or with Rights attached to an Object.
- Rights: assertions of one or more rights or permissions pertaining to an Object.

The PREMIS data dictionary defines *Semantic units*; each of them is mapped to one of the entities. (In some cases semantic units can have subunits, e.g. *identifier* can have *identifierType* and *identifierValue*, or can have *extensions* to allow the use of metadata based on external schema, e.g. *environment* has subunit *environmentExtension*).

The Data Dictionary v. 2.2 defines four entities: Object, Event, Agent and Rights in more detailed way.

6.1.1   Object Entity

Semantic unit Object aggregates information concerning digital object and describes characteristics important for preservation management. There is only one obligatory element *objectIdentifier*, which refers to all object types.

There are three types of Object:

- *Representation*: the set of files needed for a complete presentation of *Intellectual Entity*, e.g. image files of pages of a book and structural metadata;
- *File*: ordered sequence of bytes known by an operating system;
- *Bitstream*: a set of bits embedded within a file that has meaningful common properties for preservation purposes.

The above definition of bitstream differs from common usage. A detailed explanation is given in the Data Dictionary.

An Object can be linked to one or more *rightsStatements* and can take part in one or more *Events*. Following semantic units of Object are defined:

Listing 15   Semantic units of Object

```
1  objectIdentifier (M, R)
   subunits Type and Value
2  objectCategory (M, NR)
3  preservationLevel (O, R) [representation, file]
   subunits: Value, Role, Rationale and DateAssigned
4  significantProperties (O, R)
   subunits: Type, Value, Extension
5  objectCharacteristics (M, R) [file, bitstream]
   subunits: compositionLevel, Fixity
         (messageDigestAlgorithm, messageDigest,
         messageDigestOriginator), size, format
         (formatDesignation (formatName, formatVersion),
         formatRegistry (Name, Key, RoleR), formatNote),
         creatingApplication (creatingApplicationName,
         creatingApplicationVersion,
         dateCreatedByApplication,
         creatingApplicationExtension), inhibitors
         (Type, Target, Key),
         objectCharacteristicsExtension
6  originalName (O, NR) [representation, file]
7  storage (O, R) [file, bitstream]
   subunits: contentLocation (Type, Value), storageMedium
8  environment (O, R)
   subunits: environmentCharacteristic,
         environmentPurpose, environmentNote, dependency
         (Name, Identifier (Type, Value), software
         (swName, swVersion, swType, swOtherInformation,
         swDependency), hardware (hwName, hwType,
         hwOtherInformation), environmentExtension
9  signatureInformation (O, R) [file, bitstream]
   subunits: signature (signatureEncoding, signer,
         signatureMethod, signature\-Value,
         signatureValidationRules, signatureProperties,
         keyInformation), signatureInformationExtension
10 relationship (O, R)
   subunits: relationshipType, relationshipSubType,
         relatedObjectIdentification
         (relatedObjectIdentifierType,
         relatedObjectIdentifierValue,
         relatedObjectSequence),
         relatedEventIdentification
```

```
            (relatedEventIdentifierType,
            relatedEventIdentifierValue,
            relatedEventSequence)
11 linkingEventIdentifier (O, R)
   subunits: linkingEventIdentifierType,
            linkingEventIdentifierValue
12 linkingIntellectualEntityIdentifier (O, R)
   subunits: linkingIntellectualEntityIdentifierType,
            linkingIntellectualEntityIdentifierValue
13 linkingRightsStatementIdentifier (O, R)
   subunits: linkingRightsStatementIdentifierType,
            linkingRightsStatementIdentifierValue
```

Fixity is particularly important. Its subelements are: *messageDigestAlgorithm, messageDigest, messageDigestOriginator*. Typical preservation activity is validating checksum. Information about performing the test and the date should be recorded as *Event*; result of the test can be recorded as *eventOutcome*; in the *Object* should be recorded name of algorithm, e.g. *messageDigestAlgorithm* = MD5, *messageDigest* = 7c9b35da4f…and the agent as *messageDigestOriginator*.

Data Dictionary contains detailed information concerning particular elements.

6.1.2   Event Entity

The *Event* aggregates information on activities concerning one or more objects. Metadata concerning the events usually are stored outside the objects. The archive is not obliged to record all events, only important ones. Events modifying the object should always be recorded.

Elements: *eventIdentifier, eventType, eventDateTime* should always be used.

Event must be related to one or more objects and can be related to one or more agents.

Listing 16 Semantic units of *Event*
```
1 eventIdentifier (M, NR)
    eventIdentifierType (M, NR)
    eventIdentifierValue (M, NR)
2 eventType (M, NR)
3 eventDateTime (M, NR)
4 eventDetail (O, NR)
5 eventOutcomeInformation (O, R)
    eventOutcome (O, NR)
    eventOutcomeDetail (O, R)
      eventOutcomeDetailNote (O, NR)
      eventOutcomeDetailExtension (O, R)
6 linkingAgentIdentifier (O, R)
    linkingAgentIdentifierType (M, NR)
    linkingAgentIdentifierValue (M, NR)
    linkingAgentRole (O, R)
7 linkingObjectIdentifier (O, R)
    linkingObjectIdentifierType (M, NR)
    linkingObjectIdentifierValue (M, NR)
    linkingObjectRole (O, R)
```

6.1.3   Agent Entity

Semantic unit *Agent* aggregates information on attributes and characteristics of agents concerning the rights management and the management of preservation events in the archive. The only obligatory element is *agentIdentifier*. Agent can have rights and can grant rights—one or more. Agent can perform events, authorize. Agent can affect one or more objects by the events or based on (*rights statements*).

Listing 17   Semantic units of Agent

```
1 agentIdentifier (M, R)
    agentIdentifierType (M, NR)
    agentIdentifierValue (M, NR)
2 agentName (O, R)
3 agentType (O, NR)
4 agentNote (O, R)
5 agentExtension (O, R)
6 linkingEventIdentifier (O, R)
    linkingEventIdentifierType (M, NR)
    linkingEventIdentifierValue (M, NR)
7 linkingRightsStatementIdentifier (O, R)
    linkingRightsStatementIdentifierType (M, NR)
    linkingRightsStatementIdentifierValue (M, NR)
```

6.1.4   Examples of PREMIS Metadata

Data dictionary v. 1.0 contains 6 examples, showing in details how PREMIS metadata can be applied where objects have different structure:

- Microsoft Word document complete in one file,
- Electronic Theses and Dissertation (two files: a PDF and MP3),
- Newspaper complex object, Los Angeles Times (tar file that contains within a file produced by application QuarkXPress, which in turn contains links to EPS file),
- Web site (harvested Web site),
- Digital signature,
- Photograph (two files: a TIFF image and an XML file containing descriptive metadata).

For example, in case 1 Word file was ingested to archive (repository). Two digital objects were created: representation and the file. Five events were recorded: Ingest, Fixity check, Virus check, Object validation and Annotation validation.

Further examples can be found at [102].

6.1.5   Changes

PREMIS is under constant development. Procedures for requesting a change to the PREMIS Data Dictionary or to the associated XML Schema are defined at [103].

Among changes introduced to PREMIS Data Dictionary for Preservation Metadata version 3.0 are:

- Intellectual Entity became another category of Object, having the same semantic units as Representation.
- A new semantic unit *preservationLevelType* was defined to indicate preservation activities which might be applied to the object for given *preservationLevel*.
- A new semantic unit *agentVersion* was added to express the version of software *Agents*.

6.2   Besides PREMIS

The German National Library has presented in 2005 a schema for recording suitable technical metadata in the form of long-term preservation metadata for electronic resources (LMER) [104].

The project was inspired by earlier simple preservation metadata standard created by the National Library of New Zealand (2003) [105]. PREMIS did not exist at that time. The National Library of New Zealand took part in the works on PREMIS from the beginning, sharing experiences with working group.

Implementation of PREMIS can be a challenge for small or medium size libraries or archives. Some centers tried to design simpler sets of preservation metadata. An example of such approach can be found in North Carolina Preservation Metadata for Digital Objects (PMDO) [106]. System is simple—only 21 metadata elements. What is interesting, each element has mapping to PREMIS, so that institution using the PMDO can easily change it to PREMIS.

PREMIS can inspire integration with other standards. An interesting example of integrating other metadata standards with PREMIS is presented in [107]. The authors propose a model of provenance description based on integration of PREMIS OWL Ontology and ontology PROV-O.

7   Conclusion

Metadata play important role in long-term digital preservation. Metadata accompanying digital objects in current use, e.g. descriptive, technical and rights metadata, should be sent to archive together with the objects. It is recommended to gather, also from the beginning, provenance metadata concerning digital objects and send then to the archive with the objects. The Reference Model for an Open Archival Information System (OAIS) was developed to standardize digital preservation activities. Two metadata types play special role in long-term preservation: packaging metadata, applied to create packages SIP, AIP and DIP, and preservation metadata, which supports and documents process of preservation.

References

1. Digital Preservation Handbook Glossary. Digital Preservation Coalition http://www.handbook.dpconline.org/glossary. Access: 2016-10-20.
2. Digital Preservation Metadata. Angela Dappert, The British Library, Planets, PREMIS EC, Barcelona, March 2009 http://www.planets-project.eu/docs/presentations/Dappert_PreservationMetadata.pdf. Access: 2016-10-20.
3. What is Provenance. W3C http://www.w3.org/2005/Incubator/prov/wiki/What_Is_Provenance. Access: 2016-10-20.
4. Visual Resources Association http://www.vraweb.org/index.html. Access: 2016-10-20.
5. Directory Interchange Format DIF. Global Change Master Directory http://www.gcmd.nasa.gov/add/difguide/index.html. Access: 2016-10-20.
6. METS: Metadata Encoding and Transmission Standard http://www.loc.gov/standards/mets/. Access: 2016-10-20.
7. PREMIS Data Dictionary for Preservation Metadata http://www.loc.gov/standards/premis/. Access: 2016-10-20.
8. Simple Dublin Core https://www.dublincore.org/documents/1998/09/dces/. Access: 2016-10-20.
9. EBU Core metadata set for Radio archives, Tech 3293, version 1.0, 2001 http://www.tech.ebu.ch/publications/tech3293. Access: 2016-10-20.
10. European Broadcasting Union http://www.ebu.ch/home. Access: 2016-10-20.
11. Dublin Core Element Set to GCMD Directory Interchange Format DIF. Global Change Master Directory http://www.gcmd.nasa.gov/add/standards/dublin_to_dif.html. Access: 2016-10-20.
12. The Open Archives Initiative Protocol for Metadata Harvesting, v. 2.0, http://www.openarchives.org/OAI/openarchivesprotocol.html. Access: 2016-10-20.
13. Dublin Core Qualifiers http://www.dublincore.org/documents/2000/07/11/dcmes-qualifiers/. Access: 2016-10-20.
14. DCMI Metadata Terms http://www.dublincore.org/documents/dcmi-terms/. Access: 2016-10-20.
15. MARC Code List for Relators http://www.loc.gov/marc/relators/relaterm.html. Access: 2016-10-20.
16. Guidelines for implementing Dublin Core in XML http://www.dublincore.org/documents/dc-xml-guidelines/. Access: 2016-10-20.
17. XML Schema for Simple Dublin Core 2002 http://www.dublincore.org/schemas/xmls/simpledc2002/1212.xsd. Access: 2016-10-20.
18. XMLSchema for DC Terms http://www.dublincore.org/schemas/xmls/qdc/dcterms.xsd. Access: 2016-10-20.
19. arXiv.org Cornell University Library http://www.arxiv.org. Access: 2016-10-20.
20. Federacja Bibliotek Cyfrowych (FBC) http://www.fbc.pionier.net.pl. Access: 2016-10-20.
21. EBU Core Metadata Set (EBUCore). Specification 1.6. Tech 3293. Geneva, June 2015 http://www.tech.ebu.ch/docs/tech/tech3293.pdf. Access: 2016-10-20.
22. SMPTE – Society of Motion Picture and Television Engineers http://www.smpte.org. Access: 2016-10-20.
23. International Telecommunication Union http://www.itu.int/en/about/Pages/default.aspx. Access: 2016-10-20.
24. International Press Telecommunications Council (IPTC) http://www.iptc.org/. Access: 2016-10-20.
25. Metadata Guidance for the Transfer of Permanent Electronic Records. NARA Bulletin 2015-04, September 15, 2015 http://www.archives.gov/records-mgmt/bulletins/2015/2015-04.html. Access: 2016-10-20.
26. Sustainability of Digital Formats Planning for Library of Congress Collections http://www.digitalpreservation.gov/formats/. Access: 2016-10-20.

27. Revised Format Guidance for the Transfer of Permanent Electronic Records. NARA Bulletin 2014-04, February 4, 2014, http://www.archives.gov/records-mgmt/bulletins/2014/2014-04. html. Access: 2016-10-20.
28. List of Metadata Standards. Digital Curation Centre (DCC) http://www.dcc.ac.uk/resources/ metadata-standards/list. Access: 2016-10-20.
29. List of metadata standards in Wikipedia http://www.en.wikipedia.org/wiki/Metadata_ standard. Access: 2016-10-20.
30. Domain Metadata Standards. University of Central Florida Libraries http://www.guides.ucf. edu/metadata/domMetaStandards. Access: 2016-10-24.
31. Agricultural Metadata Element Set (AgMES) – AgMES 1.1 Namespace Specification http:// www.aims.fao.org/standards/agmes/namespace-specification. Access: 2016-10-20.
32. AVM – Astronomy Visualization Metadata, International Virtual Observatory Alliance standard used to tag astronomical images http://www.virtualastronomy.org/avm_metadata.php. Access: 2016-10-20.
33. PROV provenance metadata standard PROV[89] designed for interchange of provenance information http://www.w3.org/2001/sw/wiki/PROV. Access: 2016-10-20.
34. VRA Core Schemas & Documentation. Official Website http://www.loc.gov/standards/ vracore/. Access: 2016-10-24.
35. MODS: Metadata Object Description Schema http://www.loc.gov/standards/mods/. Access: 2016-10-20.
36. TEI: Text Encoding Initiative http://www.tei-c.org/index.xml. Access: 2016-10-20.
37. ONIX standards: ONIX for Books, ONIX for Subscription Products, Licencing Terms and Right Information http://www.editeur.org/8/ONIX/. Access: 2016-10-20.
38. STEP: Standard for the Exchange of Product model data http://www.steptools.com/library/ standard/. Access: 2016-10-24. – also ISO 10303: Industrial automation systems and integration – Product data representation and exchange (has many parts).
39. Metadata for Long Term Preservation of Product Data. International Symposium on XML for the Long Haul: Issues in the Long-term Preservation of XML. Proceedings August 2, 2010 http://www.balisage.net/Proceedings/vol6/html/Lubell01/BalisageVol6-Lubell01.html. Access: 2016-10-20.
40. MARC to Dublin Core Crosswalk http://www.loc.gov/marc/marc2dc-2001.html. Access: 2016-10-20.
41. Dublin Core to MARC Crosswalk http://www.loc.gov/marc/dccross.html. Access: 2016-10-20.
42. Mapping of the Dublin Core Metadata Element Set to the CIDOC CRM, Martin Doerr http:// www.cidoc-crm.org/docs/dc_to_crm_mapping.pdf. Access: 2016-10-24.
43. The CIDOC Conceptual Reference Model. International Council of Museums (ICOM) http:// www.cidoc-crm.org/official_release_cidoc.html. Access: 2016-10-24.
44. Europeana Collections http://www.europeana.eu. Access: 2016-10-20.
45. Provide data in EDM. Europeana Pro http://www.pro.europeana.eu/page/provide-data-edm. Access: 2016-10-20.
46. New railroad map of the state of Maryland, Delaware, and the District of Columbia, available at http://www.loc.gov/item/98688491/; MODS record http://www.lccn.loc.gov/98688491/ mods; Dublin Core record http://www.lccn.loc.gov/98688491/dc. Access: 2016-10-20.
47. IPTC Photo Metadata Standard (2014) http://www.iptc.org/standards/photo-metadata/iptc-standard/. Access: 2016-10-24.
48. IPTC Core and Extension Guidelines (July 2010) http://www.iptc.org/standards/photo-metadata/iptc-standard/. Access: 2016-10-24.
49. IPTC Core and Extension examples http://www.iptc.org/std/photometadata/documentation/ GenericGuidelines/index.htm#!Documents/iptccoreandextensionexamples.htm. Access: 2016-10-24.
50. Museum of History of Photography http://www.mhf.krakow.pl. Access: 2016-10-24.
51. History Meeting House http://www.dsh.waw.pl. Access: 2016-10-24.

52. Exchangeable image file format for digital still cameras: Exif Version 2.31. Standard of the Camera & Imaging Products Association. CIPA DC-008-Translation-2016. http://www.cipa. jp/std/documents/e/DC-008-Translation-2016-E.pdf. Access: 2016-10-24.
53. Exchangeable image file format for digital still cameras: Exif Version 2.2. April, 2002. Technical Standardization Committee on AV & IT Storage Systems and Equipment. Japan Electronics and Information Technology Industries Association. http://www.exiv2.org/Exif2-2. PDF. Access: 2016-10-24.
54. Exif vocabulary workspace – RDF Schema http://www.w3.org/2003/12/exif/. Access: 2016-10-24.
55. MIX – NISO Metadata for Images in XML Schema. Technical Metadata for Digital Still Images Standard http://www.loc.gov/standards/mix/. Access: 2016-10-24.
56. American National Information Standards Organization http://www.niso.org/home/. Access: 2016-10-24.
57. TIFF. Revision 6.0. Final June 3, 1992. Adobe Developers Association http://www.partners. adobe.com/public/developer/en/tiff/TIFF6.pdf. Access: 2016-10-24.
58. XMP Specification. Adding Intelligence to Media. Adobe Systems Incorporated. September 2005 http://www.partners.adobe.com/public/developer/en/xmp/sdk/XMPspecification. pdf. Access: 2016-10-24.
59. Digital Negative (DNG) Specification, Version 1.4.0.0, June 2012 http://www.images.adobe. com/content/dam/Adobe/en/products/photoshop/pdfs/dng_spec_1.4.0.0.pdf. Access: 2016-10-24.
60. International Association of Sound and Audiovisual Archives (IASA) http://www.iasa-web. org. Access: 2016-10-24.
61. IASA Cataloguing Rules http://www.iasa-web.org/cataloguing-rules. Access: 2016-10-24.
62. IASA Technical Committee, Guidelines on the Production and Preservation of Digital Audio Objects, ed. by Kevin Bradley. Second edition 2009. (= Standards, Recommended Practices and Strategies, IASA-TC 04) http://www.iasa-web.org/tc04/audio-preservation. Access: 2016-10-24.
63. ISO/IEC 13818-3:1995 Information technology "Generic coding of moving pictures and associated audio information" Part 3: Audio: http://www.iso.org/iso/iso_catalogue/catalogue_tc/ catalogue_detail.htm?csnumber=22412. Access: 2016-10-24.
64. EBU Classification Schemes and ontologies: www.ebu.ch/metadata/cs/ and www.ebu.ch/ metadata/ontologies/. Access: 2016-10-20.
65. EBU Role Code Classification Scheme http://www.ebu.ch/metadata/cs/ebu_RoleCodeCS. xml. Access: 2016-10-20.
66. FIAF – Fédération Internationale des Archives du Film / International Federation of Film Archives http://www.fiafnet.org. Access: 2016-10-24.
67. IFTA – International Federation of Television Archives[42] http://www.fiatifta.org. Access: 2016-10-24.
68. FIAF Cataloguing Rules for Film Archives http://www.fiafnet.org/images/tinyUpload/E-Resources/Commission-And-PIP-Resources/CDC-resources/FIAF_Cat_Rules.pdf. Access: 2016-10-24.
69. The FIAF Moving Image Cataloging Manual. Written by Natasha Fairbairn, Maria Assunta Pimpinelli, Thelma Ross, April 2016 http://www.fiafnet.org/images/ tinyUpload/E-Resources/Commission-And-PIP-Resources/CDC-resources/20160920% 20Fiaf%20Manual-WEB.pdf. Acess: 2016-10-24.
70. European Standards Committee (CEN) Cinematographic Works Standard (CWS) http://www. filmstandards.org/fsc.index.php/Main_Page. Access: 2016-10-24.
71. DIN EN 15744 Film identification – Minimum set of metadata for cinematographic works http://www.en-standard.eu/din-en-15744-film-identification-minimum-set-of-metadata-for-cinematographic-works/. Access: 2016-10-24.
72. DIN EN 15907 Film identification – Enhancing interoperability of metadata – Element sets and structures http://www.en-standard.eu/din-en-15907-film-identification-enhancing-interoperability-of-metadata-element-set-and-structures/. Access: 2016-10-24.

73. EN 15744 Film identification Minimum set of metadata for cinematographic works http:// www.filmstandards.org/fsc/index.php/EN_15744. Access: 2016-10-24.

74. EN 15907 Film identification – Enhancing interoperability of metadata – Element sets and structures http://www.filmstandards.org/fsc/index.php/EN_15907. Access: 2016-10-24.

75. Cox, M., Mulder, E., Tadic, L., Descriptive Metadata for Television: An End-to-End Introduction. Focal Press 2013.

76. TV-Anytime News. Schemas and Classification Schemas http://www.tech.ebu.ch/tvanytime. Access: 2016-10-24.

77. PBCore – Public Broadcasting Metadata Dictionary http://pbcore.org. Access: 2016-10-24.

78. P/META Metadata Library. EBU Tech 3295. Specification 2.2. Geneva, September 2011 http://www.tech.ebu.ch/docs/tech/tech3295v2_2.pdf. Access: 2016-10-24.

79. SPTME Metadata Registry class 13-14 – managed by Society of Motion Picture & Television Engineers (SMPTE) Registration Authority http://www.smpte-ra.org/smpte-metadata-registry. Access: 2016-10-24.

80. EBU Archives Report 2010. Technical Report 006. Geneva, June 2010 http://www.tech.ebu.ch/docs/techreports/tr006.pdf. Access: 2016-10-20.

81. Open Digital Rights Language (ODRL) http://www.w3.org/community/odrl/. Access: 2016-10-24.

82. DDEX Standards http://www.ddex.net/ddex-standards. Access: 2016-10-24.

83. METS RightsDeclarationMD Extension Schema http://www.loc.gov/standards/rights/METSRights.xsd. Access: 2016-10-24.

84. Guidelines for using PREMIS with METS for exchange, 2008 http://www.loc.gov/standards/premis/guidelines-premismets.pdf. Access: 2016-10-24.

85. XML Formatted Data Unit (XFDU). Structure and Construction Rules. Recommended standard. CCSDS 661.0-B-1. Blue Book, September 2008. ISO equivalent 13527:2010 http://www.public.ccsds.org/Pubs/661x0b1.pdf. Access: 2016-10-24.

86. LOTAR: Long Term Archiving and Retrieval. Metadata for Archival Package Workgroup http://www.lotar-international.org/lotar-workgroups/metadata-for-archival-package.html. Access: 2016-10-24.

87. E-ARK. WP4 Archival Records Preservation. Work Packages. http://www.eark-project.com/about/work-packages/9-about/32-wp4intro. Access: 2016-10-24.

88. METS Implementation Registry http://www.loc.gov/standards/mets/mets-registry.html. Access: 2016-10-24.

89. METS registered profiles http://www.loc.gov/standards/mets/mets-registered-profiles.html. Access: 2016-10-24.

90. CCSDS Missions http://www.public.ccsds.org/implementations/missions.aspx. Access: 2016-10-24.

91. Project CASPAR – Cultural, Artistic and Scientific knowledge for Preservation, Access and Retrieval – an EU Integrated Project http://www.cordis.europa.eu/project/rcn/92920_en.html. Access: 2016-10-24.

92. Project SAFE – Standard Archive Format for Europe http://www.earth.esa.int/SAFE/. Access: 2016-10-24.

93. Universal Object Format An archiving and exchange format for digital objects. Project kopal Co-operative Development of a Long-Term Digital Information Archiv. By Dipl.-Inform. Tobias Steinke Frankfurt am Main 2006 http://www.kopal.langzeitarchivierung.de/index_objektspezifikation.php.en. Access: 2016-10-24.

94. Caplan, P., Repository to Repository Transfer of Enriched Archival Information Packages. D-Lib Magazine, November/December 2008, vol. 14, number 11/12. http://www.dlib.org/dlib/november08/caplan/11caplan.html. Access: 2016-10-20.

95. METS. Metadata Encoding and Transmission Standard: Primer and Reference Manual. v. 1.6, 2010, Digital Library Federation http://www.loc.gov/standards/mets/mets-schemadocs.html. Access: 2016-10-24.

96. METS Example Documents http://www.loc.gov/standards/mets/mets-examples.html. Access: 2016-10-24.

97. Producer-Archive Interface Methodology Abstract Standard (PAIMAS). Recommendation for Space System Practices. CCSDS 651.0-M-1. Magenta Book. May 2004. ISO equivalent 20652:2006 http://www.public.ccsds.org/Pubs/651x0m1.pdf. Access: 2016-10-24.

98. Producer-Archive Interface Specification (PAIS). Recommended Standard. Blue Book. Issue 1. February 2014. CCSDS 651.1-B-1. ISO equivalent 20104:2015 http://www.public. ccsds.org/Pubs/651x1b1.pdf. Access: 2016-10-24.

99. XML-binary Optimized Packaging, W3C Recommendation, 25 January 2005 http://www. w3.org/TR/xop10/. Access: 2016-10-24.

100. Preservation Metadata: Implementation Strategies http://en.wikipedia.org/wiki/Preservation_ Metadata:_Implementation_Strategies. Access: 2016-10-24.

101. PREMIS Data Dictionary for Preservation Metadata. v. 2.2, July 2012, PREMIS Editorial Committee http://www.loc.gov/standards/premis/v2/premis-2-2.pdf. Access: 2016-10-24.

102. PREMIS examples http://www.loc.gov/standards/premis/examples.html. Access: 2016-10-24.

103. Procedures for requesting a change to the PREMIS Data Dictionary or to the associated XML Schema http://www.loc.gov/standards/premis/revision-process.html. Access: 2016-10-24.

104. Long-term preservation Metadata for Electronic Resources (v. 1.2) http://www.dnb.de/EN/ Standardisierung/LMER/lmer_node.html. Access: 2016-10-24.

105. National Library of New Zealand. Preservation Metadata. Metadata Standards Framework Metadata Implementation Schema. July 2003 http://www.digitalpreservation.natlib.govt.nz/ assets/Uploads/nlnz-data-model-final.pdf. Access: 2016-10-24.

106. NC Preservation Metadata for Digital Objects. 2013 EDITION BY LISA GREGORY, KATHLEEN KENNEY, AMY RUDERSDORF http://www.digitalpreservation.ncdcr.gov/ pmdo2013final.pdf. Access: 2016-10-24.

107. Provenance Description of Metadata using PROV with PREMIS for Long-term Use of Metadata, Chunqiu Li, Shigeo Sugimoto. DC-2014 in Austin, Texas, 8-11 October 2014 http:// www.dcpapers.dublincore.org/pubs/article/view/3709. Access: 2016-10-24.

Part II
Solutions Proposed by the CREDO Project

The CREDO Project

Tomasz Traczyk and Włodzimierz Ogryczak

Abstract The chapter describes main assumptions of the project *Digital Document Repository CREDO*, which resulted in design and implementation of a long-term digital archive of significant capacity, conforming to OAIS standard, with packaging preserved resources in archival packages, and metadata management. The archive ensures storage correctness and devices reliability monitoring. Thanks to sophisticated scheduling of archive operations, energy efficiency is also provided. A discussion is presented how the CREDO archive meets the requirements, which are usually applied to long-term digital archives.

1 Introduction

The project entitled *Digital Document Repository CREDO*[1] was submitted as a part of the pilot undertaking of the Polish National Centre for Research and Development, entitled '*Demonstrator* + Supporting scientific research and development works for demonstration scale' [5].

In Latin *credo* means 'I believe', which seems to be quite a good watchword for trustworthy digital repository.

The CREDO project was carried out by the consortium comprising Polish Security Printing Works [6] (consortium leader), Warsaw University of Technology [9], and Skytechnology Ltd. [7].

The goal of the project was to design and launch a demonstrational version of a digital repository enabling short- and long-term archiving of large volumes of

---

[1]CREDO is the acronym of Polish name *Cyfrowe REpozytorium DOkumentów*, which means 'Digital Document Repository'.

T. Traczyk (✉) · W. Ogryczak
Institute of Control and Computation Engineering, Warsaw University of Technology,
Warsaw, Poland
e-mail: T.Traczyk@ia.pw.edu.pl

W. Ogryczak
e-mail: W.Ogryczak@ia.pw.edu.pl

© Springer International Publishing AG 2017
T. Traczyk et al. (eds.), *Digital Preservation: Putting It to Work*,
Studies in Computational Intelligence 700, DOI 10.1007/978-3-319-51801-5_3

digital resources. By design the repository is to act both as a secure file storage and as a digital archive providing metadata management, archival packages, etc.

One of the primary functions of the system is to support various currently available data carriers: hard drives, solid state drives, tapes. The repository ensures a high level of security of the stored information. Reliability of information readouts is ensured by data replication mechanisms in the used filesystem, as well as by distributed nature of the system that will enable storing copies of the resources in more than one location. The repository architecture enables replacement and continuous upgrades of individual system components (subsystems, filesystems etc.) following the emergence of new technologies.

The solution is designed primarily for institutions, which store large digital resources for long periods of time, e.g. cultural institutions, mass media, state administration offices, health care institutions, etc.

2 Assumptions of the Project

The CREDO project has two main goals:

- Creating a short-term on-line digital repository, based on reliable file system with data replication, having petabyte capacity, that could store files of multi-terabyte size.

  In this repository the management of stored resources (e.g. checking their completeness, metadata management, etc.) is on the user side. The management of storage devices and their reliability, as well as of proper file replication and replicas integrity is on the repository side.

  The main difficulty lies in very large volume of the whole data, and in enormous size of single files.

- Design and implementation of a long-term digital archive, which can ensure durability and usability of stored resources on horizon of dozens or even hundreds of years. The archive must guarantee credible restoring of the resources, metadata management and search, and many other features commonly expected from digital archive.

  To meet these expectations, the archive should conform to recognized standards, and should be able to pass certification. The management of the stored resources, as well as of the processes executed in the archive, should of course be on the archive side.

  The main difficulty, in this case, lies in the very long time of expected operation of the archive, which is substantially longer that longevity of any digital technology or popularity of any digital data format.

The additional project goal is to build a demonstrative installation of 2 PB volume, dislocated in two distant locations.

2.1   CREDO Long-Term Digital Archive

The most interesting part of the project, which is also the main subject of this book, is a design of a long-term digital archive. This archive should provide:

- long-term persistence of the stored digital resources;
- guarantee of credible restoring the data;
- data dislocation in distant data centers;
- complete functions of the digital archive, including:

  - conformance to OAIS [1] and other archival standards,
  - packaging the resources together with their metadata in archival packages,
  - metadata management and effective search;

- energy efficiency of the storage;
- readiness to be certified;
- ability to establish co-operation between different archives.

3   The CREDO Project Versus Requirements for Digital Preservation

3.1   Information Longevity

It can be expected that the longevity of the information should be ensured by long guaranteed durability time of the data carriers.

Unfortunately, so far no such digital data carriers exist, which can guarantee reliable data preservation for hundreds or even many dozens of years. What can we do is to copy the information to many (possibly distant) places and to periodically move it from older carrier to a newer one. In CREDO we decided to use standard data carriers and to add longevity by proper and automatized procedures of data duplicating and rewriting.

3.1.1   Data Carriers in CREDO

Currently, the most popular type of data carrier for information archival is probably LTO tape. Storing data on the tapes is relatively cheap, but there are some important disadvantages of this type of storage:

- the access to the data is sequential, so rather inconvenient;
- all tapes should be rewound (retensioned) at regular intervals, usually not less than every three years [8];

- the data should be periodically read and written to refresh the magnetic signal and prevent data loss [4, 8] while, due to the sequential access, this operation is difficult and time-consuming;
- though successive generations of LTO tapes has compatible cartridges, LTO drives can usually read only their own generation and two prior generations, so it is possible that some older tapes may not be readable if older drives spoil.

Due to all these problems, in the CREDO project magnetic tapes are considered only as an alternative storage type, and the attention is focused on hard-disk based storage. The current CREDO storage is based mainly on magnetic disks, with a minor part composed of SSD drives. Full transition to SSD drives is however straightforward. In the future it is possible to use other data carriers, and the architecture of the system should ensure that such a transition will be quite easy.

An open architecture of CREDO enables easy adoption of new software, optimizing use of new data carriers, implementing new strategies of energy saving, new reliability checking algorithms, additional data protection methods, etc. It is also quite easy to automatize the migration from older data carriers to new ones.

3.1.2 Reliable Resources Persistence

Reliability of the information persistence in CREDO is based on several factors:

- data redundancy and dislocation,
- data integrity monitoring,
- periodical magnetic refreshment,
- carriers reliability monitoring,
- data relocation from not reliable areas and optimization of data allocation.

Data Redundancy and Dislocation

In the CREDO archive a two-level data redundancy is implemented to ensure data persistence.

Low-level data redundancy is built into used SZPAK filesystem (see chapter "CREDO Repository Architecture"). This method bases on low-level data chunk duplication, which is managed internally by the filesystem, and is transparent to its users. The number of these low-level copies can be, however, managed by the archive by setting appropriate extended attribute of the file or directory. In CREDO we recommend not less than two low-level copies if the resource has also a high-level replica, and not less than three low-level copies otherwise.

In future versions of CREDO other low-level redundancy mechanisms may be implemented, e.g. error-correcting codes.

High-level redundancy is managed by the CREDO archive. It bases on storing the whole archival package (AIP) in more than one replica. The replicas can be stored in separate areas of the filesystem (e.g. using different storage technologies) or in separate filesystems. This high-level replication enables data dislocation, as the replicas may be stored in physically distant parts of the archive.

In CREDO we recommend storing the resources in two replicas, placed in distant (dislocated) archive areas.

Higher level of the data dislocation is possible within the federation of archives. Dislocated replicas can be stored in several archives, with mutual awareness of having the replicas and of the state of their correctness. Some coordination of activities related to copies is possible, especially those activities, which encounter any risk of damage (e.g. package migration). In such higher-level dislocation, the replicas may not only be technologically distinct, but even can have not bit-identical, although semantically equivalent content. Some ideas concerning this type of replication have been developed in the CREDO project, as described in chapter "Information Management in Federated Digital Archives".

Data Integrity Monitoring

As digital storage carriers are not perfect, careful data monitoring is necessary. It is necessary also because of the certification requirements.

All files stored in the CREDO archive are periodically checked for their compliance with declared digital digests. These digests are stored in several copies: as the extended attribute of each file, in metadata files stored in AIPs, and in the archive database. The digital digests loss or distortion is therefore very unlikely. In current version of the CREDO archive SHA256 digital digests are used, but it is possible to use other digest type, or even more than one type.

This data integrity check demands periodical powering-on the disks, which is also necessary for maintaining the good mechanical condition of the devices.

Periodical Magnetic Refreshment

The signal recorded on magnetic carriers must be refreshed from time to time, because of magnetic particle instabilities and related data loss risk [8]. In CREDO the magnetic refreshment is periodically requested by the archive and low-level executed by the filesystem.

Data Carriers Reliability Monitoring

The reliability of the data carriers should be monitored. In the CREDO archive this monitoring is based on reading and analyzing S.M.A.R.T. data from the disks. More details can be found in chapter "Persistence Management in Long-Term Digital Archive".

The architecture of CREDO enables easy addition of new reliability monitoring methods for other types of data carriers.

Data Relocation from Not Reliable Areas and Optimization
of Data Allocation

If any of archival storage areas are considered not enough reliable, the data should
be migrated to more reliable areas. In the CREDO archive this relocation is managed
automatically: the archive sends to the filesystem appropriate requests to remove
unreliable carriers, and the filesystem migrates the data before disconnecting these
carriers.

When new data is allocated in the archival storage, its allocation is optimized.
Allocation in more reliable storage areas is preferred.

3.1.3   Transition to New Technologies

As the archive should last dozens of years, it is obvious that the technologies used to
store data will change. Therefore, it is crucial to the data longevity, that the archive
should be able to adopt emerging technologies, and integrate new storage methods
with earlier ones.

In CREDO we assume relative independence of current technologies. The archive
is able to work with many filesystems, which may be implemented in different technologies. High-level replicas may be placed in distinct filesystems with technological
diversification. Also the data relocation may be carried out between the data areas
implemented in different technologies, so the migration from older to newer technologies may be almost seamless.

In the current version of CREDO, filesystem compliance to the POSIX/SUS standard is assumed. This compliance may be, however, obtained also for non-POSIX
filesystems, by creating additional layer between the real filesystem and CREDO.

3.2   Verifiability of Information Correctness

Verifiability of the stored information correctness is crucial to ensure the information
persistence. This correctness means information integrity and authenticity. Hence,
methods must be provided to verify these two conditions.

3.2.1   Information Authenticity

Information authenticity means that no damage or unauthorized change of the information happened. In CREDO, periodical two-level data monitoring is assured, as
described in Sect. 3.1.2. As the list of each package files, along with their digital
digests, is stored not only in the archival storage, but also in the archive database, the
completeness, and permanence of each archival package can be reliably verified.

3.2.2 Information Integrity

Information integrity means its completeness and conformity of the content with its declaration (also in terms of the requirements of the data formats used). In order to check the integrity, the metadata of each archival package must be used, which describe declared content, or at least the initial/original state of the package. Storing metadata in the archive is therefore essential for the proper archive operation.

In CREDO, the archived resources are packed into archival packages together with their metadata. The metadata are stored in XML files. Textual, simple and self-descriptive XML format ensures the ability of quite easy and correct metadata interpretation after years.

Copies of the selected metadata are stored in the archive database, parsed into database structures or as XML documents. Generic database structure enables the system to use of various metadata standards, also no yet existent. The XML form of metadata storage is even more flexible.

Mechanisms are implemented to check the integrity of archival packages. Package completeness, i.e. compliance of the contents of the package with its declaration (or initial state) is verified on a regular basis, and after each package modification or transition. Correctness of file formats and their conformity with format declarations may also be checked, usually during ingest of the package.

To make the information integrity verifiable over the lifetime of the digital resource, the resource supplier should prepare and deliver the metadata in SIP package. This metadata should contain at least the list of delivered files and their digital digests. As the list may be delivered and confirmed in a formal way, a non-repudiation of the archived information may be ensured. Declarations of file formats are recommended. Conformance of actually delivered data to these declarations can be verified by the archive.

If the metadata cannot be delivered by the resource supplier, CREDO itself can generate replacement metadata reflecting the initial state of the delivered information. More details on metadata processing can be found in chapter "Metadata in CREDO Long-Term Archive".

Metadata must be delivered using agreed standards. In current CREDO version a limited version of METS standard is adopted, as this standard is commonly used by Polish archives and libraries. It is possible and quite easy to use other standards in the future, e.g. XFDU.

The process of information delivery should be based on clearly defined responsibilities of suppliers and the archive, and well-defined procedures conforming to widely recognized standards. Strict procedures and compliance with standards allow efficient ingest of large amounts of information by the archive. Strict procedures can also facilitate the settlement of any dispute regarding the information authenticity and integrity.

3.3 Availability of the Information

Availability of the information requires its easy finding, effective obtaining and possibility of its proper interpretation.

3.3.1 Acquisition of the Information from the Archive

The possibility of finding and obtaining the requested resource is ensured by the archive, by its search and outgest services.

The information can be effectively searched by querying the archive database. This process requires no access to archival packages preserved in (usually off-line) archival storage. Through efficient search mechanisms existing in the database, the search can be relatively fast.

There are many criteria, which may be used in a query to the CREDO archive: basic file parameters (file name, creation date, creator etc.), multiple types of resource identifiers (DOI, URI, etc.), descriptive metadata (e.g. conforming to Dublin Core standard). The list of the criteria is not closed, and it may quite easily be expanded by adding new metadata elements, criteria types and search methods.

The metadata may be delivered by resource suppliers right in SIP packages (e.g. in delivered METS files). They may also be extracted from many popular file formats, from so-called embedded metadata (e.g. EXIF, IPTC or XMP metadata in graphic files). Recovered descriptive metadata may next be mapped to Dublin Core standard, to facilitate query making.

Though it is necessary to provide some resource search methods, a deep interpretation of metadata is not the task of a long-term archive. Basic metadata should, however, be obtained to the archival database, possibly unified for efficient search.

More complex metadata (e.q. extracted embedded metadata) can also be stored in the database in XML form, which allows complex searches using XQuery language. In many cases embedded metadata may be hard to bring to simple data structures, but their XML representation is usually quite straightforward.

3.3.2 Information Interpretability

Interpretability of archived resources is one of the most difficult issues in long-term digital archiving, because of the fast moral aging of digital data formats and possible atrophy of so-called designated community, i.e. a group of people, who can properly understand and use the restored information.

The latter problem cannot, of course, be addressed by any technical or even organizational means. Problem of outdated formats can, however, be at least partially solved by preserving documentation of the formats. The actual use of these formats after years can however still be difficult, because of the lack of appropriate software tools. This leads to the conclusion, that only selected formats are well-suited for archival purposes. This problem is discussed in chapter "Requirements for Digital Preservation".

Though digital data in any format can be stored in the CREDO archive, only selected formats, assuring correct interpretation of resources over long time horizon, are recommended. A full range of CREDO services (e.g. embedded metadata extraction) is provided only for recommended formats. The CREDO system can even reject files not conforming to format recommendations, in accordance with the agreement with aresource supplier. For the recommended file formats, documentation of the formats can be stored in the archive and related to the resources, which use these formats.

3.4 Confidentiality of the Information

The information stored in the archive should be available only for authorized recipients.

In the CREDO archive a physical and technical protection is provided in accordance with the highest industry standards, as the leader of the CREDO project is engaged in production of securities, bills, passports, etc.

An access to the CREDO system is possible only via protected VPN. Archive users never obtain any direct access to archival storage. For most demanding archive customers, the possibility exists to dedicate separate protected filesystems.

CREDO programs gain direct access to the archival storage in a limited and controlled way. Dedicated CREDO security subsystem is used to authorize other CREDO programs to operate on archive files, always with the smallest sufficient privileges, and only for the time necessary to execute their operations. All actions are registered in system journals (logs).

In the current version of CREDO, a model of access permissions is very restrictive. The resources stored in the archive are available only for their owner/supplier and users registered by the owner. Even resource metadata is available only for resource owner and subjects authorized by the owner. This restrictions can of course be replaced in the future by relaxed, more complex authorization model.

3.5 Economical Effectiveness of the Storage

Any digital archive must have acceptable operation costs. The cost of energy is one of the major costs of each digital archive operation, especially if based on disk storage. The cost of energy can however be significantly reduced by an appropriate archive management: storage media and servers should be switched off when they are not needed.

Energy efficiency is one of main goals of the CREDO project. To make its large volume of storage devices energy efficient it is necessary to power off most of devices for most of the archive operation time. This leads to so-called deep (off-line) archive, which offers 'on order' rather than 'on demand' access model. Requests to the archive are queried and their execution time is optimized to minimize power consumption. Advanced optimization algorithms control 'intelligent' powering on and off the servers and storage devices. Access to the archival storage should be scheduled to minimize the time of storage devices operation, but maintaining acceptable data access time. The scheduling algorithms should however provide sufficiently frequent powering of data carriers, due to the requirements of specific technology and to ensure the timely execution of maintenance (e.g. refreshment, relocation) operations.

Disk-based digital archive without intelligent power management generates unacceptable energy costs. Effective energy management requires a fairly sophisticated scheduling algorithms—it is a challenge even for experienced experts. A broad description of used optimization methods can be found in chapter "Power Efficiency and Scheduling Access to the Archive".

3.6 Standardization and Certification

3.6.1 Standards in Digital Archive

Only compliance with widely recognized standards can ensure a possibility to correct interpretation of archived resources on a long-term horizon. These standards should relate to the content of the archive: formats of stored resources and metadata. The archive structure and procedures should also conform to the standards or commonly accepted good practices.

In CREDO compliance of the archive philosophy and activities with the Open Archival Information System (OAIS, ISO 14721:2012) [1, 3] recommendation is assumed. This standard specifies a reference model for digital archives.

The use of standard metadata formats (METS, PREMIS) provides possibility to exchange the metadata with many organizations and increases the chances of proper interpretation of the information in distant future. Also file formats recommended (or enforced) by the CREDO archive are mostly compliant with recognized open standards.

3.6.2 Certification of the Archive

Confidence in the digital archive should be based on its certification. OAIS reference model [1] provides the terminology and logical structure for the certification, and its rules are specified in [2].

In CREDO the archive ability to be certified is assumed. It is supported by compliance with the OAIS model, clear archive architecture with well-defined assignment of tasks to archive subsystems, detailed logging of actions, events and errors, etc. Technical documentation of the system is provided in English due to the potential lack of local certification bodies.

4 CREDO Implementation

In the CREDO project an implementation of the digital archive in so-called demonstration scale [5] was assumed. This level of implementation is significantly higher than the prototype, but is not yet ready for work in production environment. Current version of the system can however be quite easily and quickly adapted to the production requirements, as the hardware is complete, running and placed in secure locations, the archival storage is quite large (over 2 PB) and ready to be extended, and the software can be quite easily adapted to the needs of actual customers.

CREDO Filesystem

CREDO archive can potentially use various filesystems compliant with POSIX standard. For full use of the advantages of CREDO an adapted filesystem is however necessary. To enable energy efficiency, the filesystem should be able to allocate the data in separately switched areas. To make data carriers monitoring possible, the filesystem or some its extension must be able to report data carriers state, e.g. S.M.A.R.T. data of the disks.

In the current version of CREDO a distributed file system SZPAK is used, which is based on MooseFS open-source project and adapted to the needs of CREDO archive. Three instances of SZPAK filesystem are utilized: one for archive buffer, and two for archival storage, placed in two distant locations. More detailed description can be found in chapter "CREDO Repository Architecture".

The CREDO archive can quite easily utilize any filesystem compatible with POSIX, possibly with some added functionality, as stated above.

Data Carriers in CREDO

The CREDO storage is based mainly on disks, for less difficulties in refresh and relocation, and for the possibility to build an on-line repository in the same technology (which was one of the project goals). The solution based on LTO tapes is however also built in the SZPAK filesystem.

It is assumed that CREDO storage should be constructed of typical components available for reasonable prices. Popular disks with good capacity/price ratio, and typical LTO tapes are used. It is likely that, along with falling prices of SSD disks, the greater part of the archival storage will base on this type of carriers. In the future CREDO can also support other, new media types.

5 Conclusion

In the CREDO project a working digital archive of significant capacity has been created, which operates in accordance with OAIS, works with good economic efficiency, and is fit for certification.

Two distant co-operating repositories with a total capacity of more than 2 PB have been built. The key features of a digital archive conforming to OAIS, with archive sessions and packaging stored resources and metadata into archival packages have been implemented. The system is open to new technologies, with great potential for further development and customization to the needs of users.

The most interesting features of the CREDO archive include:

- a flexible model of archival session control, suitable to co-operate with optimization algorithms of reasonable complexity;
- a complex and interesting method of scheduling, aimed at saving energy used by the archive;
- methods to control the allocation of files at the assumed incomplete access to information from the file system (in so-called open-loop system);
- generic data structures, which are able to store various metadata, also user-defined or defined by future standards;
- XML storage of metadata, which is extremely flexible and enables very complex queries;
- flexible, but quite simple methods of selective embedded metadata extraction and mapping to given standard (e.g. Dublin Core).

The experiences from the project can be used in further and new work, especially in cooperation with actual users of digital archives.

The CREDO repository provides data storage with high reliability (bitstream preservation), and includes OAIS-compliant services to ensure long-term e.g. of archived content (content preservation) at reasonable cost.

References

1. Consultative Committee for Space Data Systems: Reference model for an open archival information system (OAIS). Recommended practice. (2012). URL http://www.public.ccsds.org/pubs/650x0m2.pdf. Access: 2016-10-25.
2. International Standard Organization: Space data and information transfer systems – audit and certification of trustworthy digital repositories ISO 16363:2012 (2012). URL http://www.iso.org/iso/catalogue_detail.htm?csnumber=56510.
3. International Standard Organization: Space data and information transfer systems – open archival information system (oais) – reference model (2012). URL http://www.iso.org/iso/catalogue_detail.htm?csnumber=57284.
4. National Archives and Record Administration: Standards for guidance on maintenance and storage of electronic records: Part 1234 – electronic records management (2006).
5. National Centre for Research and Development: Demonstrator+. URL http://www.ncbir.pl/en/domestic-programmes/demonstrator. Access: 2016-10-25.
6. Polish Security Printing Works (Polska Wytwórnia Papierów Wartościowych). URL http://www.pwpw.pl/. Access: 2016-10-25.
7. Skytechnology Ltd. (2016). URL http://www.skytechnology.pl. Access: 2016-10-25.
8. Van Bogart, J.W.: Magnetic Tape Storage and Handling: A Guide for Libraries and Archives. Commission on Preservation and Access and the National Media Laboratory (1995). ISBN 1-887334-40-8.
9. Warsaw University of Technology (Politechnika Warszawska). URL http://www.pw.edu.pl/. Access: 2016-10-25.

CREDO Repository Architecture

Tomasz Traczyk

Abstract A long-term archive has to be designed in such a way that it can function for decades, despite changes in technology and methods of storage. As the technology rapidly changes, we cannot assume any currently designed module of the archive to be permanent. Therefore, an architecture of the archive has to enable quite easy and safe exchange of system modules and smooth cooperation between older and newer solutions and technologies. The architecture itself should be as stable as possible. The chapter presents assumptions and some details of the architecture of the CREDO archive. Subsystems of the archive are described and currently used technologies are briefly discussed.

1 Introduction

The architecture of a long-lasting information system must be designed with a special care, as the architecture itself should be stable over dozens of years, but the system components must be able to be exchanged.

In the case of long-term digital archive, the proper architecture is particularly important, because it is certain that the technology of data storage will rapidly evolve, so frequent changes of the system components will be required.

Therefore, the architecture has to promote constant development and exchange of components and technologies, at reasonable cost and low risk.

2 Main Assumptions of the CREDO Architecture

Main CREDO technical objectives include:

- reliable and cost effective persistence of digital resources,

T. Traczyk (✉)
Institute of Control and Computation Engineering,
Warsaw University of Technology, Warsaw, Poland
e-mail: T.Traczyk@ia.pw.edu.pl

© Springer International Publishing AG 2017
T. Traczyk et al. (eds.), *Digital Preservation: Putting It to Work*,
Studies in Computational Intelligence 700, DOI 10.1007/978-3-319-51801-5_4

- easy adaptation to changes of technology,
- focus on sustainable development.

These objectives are to be achieved through the following means:

- data redundancy and storage media refreshing,
- storage monitoring and relocation of data from less to more reliable storage areas,
- repository distribution and data dislocation,
- system modularity,
- exchangeability of used technologies and system components,
- in particular, exchangeability of media technology and filesystems.

The most important architectural assumption of CREDO is to achieve the highest possible long-term independence of the system from used technology. For the horizon of dozens of years, we cannot expect durability of any information technology. Therefore the longevity of the system requires exchangeability of each used technology.

2.1 Means to Achieve Long-Lasting Solutions

Modularity of the CREDO System

The CREDO system is divided into loosely coupled subsystems. Interfaces between them are well-defined, so it is possible and relatively easy to exchange an individual subsystem without affecting others.

Interoperability with external systems, e.g. user/operator financial systems can be achieved by means of open interfaces based on XML messages, e.g. REST Web services.

The whole archival system may comprise a plurality of federated, but independent archives with dislocation of stored resources between them.

Replaceable Subsystems and Layers

CREDO subsystems are loosely coupled, so that they can be replaced independently. Software layers are designed in accordance with OAIS recommendation [1, Annex E], which proposes five software layers:

- media layer;
- stream layer, which deals with named bit streams, e.g. files;
- structure layer, which organizes the bit stream into primitive data structures;
- object layer, which organizes the data structures into objects having well-defined semantics;
- application layer.

Due to loose coupling between the layers (especially between stream layer and higher ones), multiple filesystems, built in various technologies, can coexist in one CREDO archive.

Adaptation to Technology Changes

Long-lasting information system should be relatively independent of current technical solutions, as computer-related technologies change rapidly, and often these changes are unpredictable. This independence can be achieved by system modularity and exchangeability of modules together with their implementation technology. Development of the system is facilitated by compliance with recognized standards, weekly related to any technology, like OAIS and XML. In particular, system extension to operate on new filesystems, using various storage media (also not yet existent) must be relatively easy, as the archive should be able to quickly adopt new storage technologies.

Focus on Sustainable Development

As the long-term digital archive should operate for decades, its hardware and software should be designed and maintained with the assumption of continuous development. This should include relatively easy extension with new types of media and/or filesystems, exchangeability of whole subsystems achieved through loose couplings and well-defined interfaces, flexible data structures, and relatively easy addition of new subsystems.

The archive should be scalable, i.e. it should enable scaling of filesystems used so far, but also quite easy addition of new instances of filesystems, and new data centers.

The used software should be portable, i.e. able to be moved to new hardware and/or operating systems. Archive management in CREDO is based on technologies with very high portability (Oracle, Java), and the CREDO filesystems are based on open-source technologies, having the ability to adapt to new platforms without dependence on proprietary solutions, licensing, etc.

Data structures and procedures for information processing should be designed according to the principle of application independence, i.e. so that they can be used in various processes, including not yet recognized.

3 CREDO Technologies

3.1 Technical Standards in CREDO

Compliance with recognized standards is one of the main objectives of the CREDO project. The archive structure and operation conform to OAIS recommendation [1], which is recognized as the most important standard for digital archives.

Communication between subsystems, and metadata storage and exchange bases on XML documents. CREDO widely uses XML-related standards: XML [2], XML Schema [3], XML Namespaces [4], and XQuery [5]. XML dialects: METS [6], and PREMIS [7] are used for metadata storage and exchange.

Filesystems used by CREDO must be compliant with POSIX/SUS [8] standard.

3.2 The System Architecture and the Technology Independence

The long-term archive should be able to be effectively used in period of several dozens or even hundreds of years. It is therefore not possible to implement its functions in a manner, which will not have to be changed during operation of the system, because of rapid advances in technology and fast "moral" aging of IT solutions.

That is why it is very important to organize operation of the archive so that it can function possibly regardless of any specific technology:

- procedures of the archive should not be dependent on specific technology, and should be as stable and invariable as possible;
- system architecture should enable and facilitate exchange of individual components (e.g. subsystems) for new versions, probably implemented in new technologies, without disturbing the operation of other components and the whole archive;
- technical means, which gives hope for long-term stability or can be easily replaced, should be chosen;
- non-standard, niche or long-term impermanent technologies should be avoided in order to avoid system modernization in relatively near future.

The current implementation of CREDO obviously have to use currently available, specific technologies and tools. In the design process, however, the use should be avoided of very specific features of particular tools and unconventional or bizarre solutions, e.g. unusual data structures or non-standard database features. This approach facilitates possible future reimplementation of system modules in new technologies.

Software implementation of CREDO functions is built with the following principles:

- modules are as simple as possible, i.e. realize the smallest, possible to separate, parts of the functionality;
- the functions of the individual modules are precisely defined and documented;
- dependencies between modules are as weak as possible;
- modules use well-described common data structures in the database;
- modules should separate their specification from the implementation;
- the APIs of the modules should be well-documented and should not be subject to change.

Compliance with these rules ensures relatively easy replacement of individual CREDO subsystems, as communication between the subsystems is well defined, assumed not changing, and possibly weakly dependent on the technology. Of course, technical details of calling the services of the modules may be dissimilar in various technologies, but at the logical level (parameter lists, etc.), the interaction between the modules should not depend on technology.

It is necessary to precisely define the tasks of the subsystems, layers and modules, to establish clear interfaces between the subsystems and use of flexible, and therefore stable, data structures (in XML messages/documents and in the database).

In CREDO the algorithms and programs are designed to be as simple as possible. It is achieved through proper system architecture, flexible data structures and using— whenever it is possible and efficient—data-driven programs, which algorithms are not hard-coded into the software, but stored in appropriate data structures in the database.

3.3   Persistent and Replaceable Components

Since the longevity of the system requires certain persistent elements, it is necessary to distinguish the technical components of CREDO to permanent and replaceable ones. Persistent components are selected so that the need to change them is unlikely on a long time horizon. They can at most evolve, with maintaining backwards compatibility. Persistent components include:

- used standards,
- system architecture,
- interfaces (including message formats),
- structure of the AIP package,
- procedures of the archive,
- recommendations for formats, metadata, etc.

Replaceable components are those for which we cannot assume that they will not be subject to change in the foreseeable future. These components may be replaced or complemented by new technologies, and old and new solutions can coexist. Such components include:

- hardware and operating systems,
- data carriers and supporting software (drivers, etc.),
- filesystems,
- specific IT solutions and tools (database management system, programming languages, etc.) used to develop the CREDO subsystems,
- communication protocols,
- structures of SIP and DIP packages,
- data formats.

3.4   OAIS Layered System Model in CREDO

The OAIS standard [1] recommends that over the media layer, i.e. the data carriers and the equipment for their reading and writing, there is so-called stream layer, which delivers streams of bytes under certain names. Such a layer may be provided by standard filesystem, supplemented with some additional functions necessary for the long-term repository. This approach was adopted in the CREDO project.

The media layer in CREDO is therefore covered by a stream layer, which allows to change data carriers without interfering with the rest of the system. Functions of the layers and the communication between them should be designed in a way, which enable the layer be easily replaced by a new, different implementation. In CREDO it is also assumed that different implementations of stream layer should be able to simultaneously coexist in the system.

The stream layer of the CREDO repository performs the functions of POSIX/SUS (Single UNIX Specification) [8] compliant filesystem, supplemented with features related to control of memory allocation, media reliability, and power management, as described in Sect. 3.5.

The functionality of the repository is built by creating an appropriate software (corresponding to structure, object and application layers defined in OAIS recommendation) over the stream layer.

Content Versus Bitstream Preservation in CREDO

The long-term archive CREDO provides all these layers and a "content preservation" level of persistence for metadata. For digital resources themselves, CREDO provides essentially only the media and stream layers, so a "bitstream preservation" level of persistence is directly achieved. In some functions of the archive (e.g. reading the metadata from the archived files) more layers, however, are partially implemented.

By managing metadata and information about the data/file formats and—which is recommended—by storing the documentation of these formats in the archive, the level of resources persistence can be raised close to "content preservation". Factors missing to fully ensure this level of persistence are mainly not technical, for example related to management of so called designated community (as defined in OAIS), and can be completed in the archive organizational rather then technical environment.

The only technical means, which lacks for "content preservation" level, is migration of stored resources to new formats. This functionality can however be added in future versions of CREDO. Due to flexible data-driven method of archival sessions definition and execution, extending the archive with such a new functionality is relatively easy.

3.5   Filesystem Technology

In the current version of CREDO, a dedicated filesystem SZPAK, prepared by project partner Sky Technology [9], is used. It is a clone of the recognized open-source project MooseFS [10], adapted to the requirements of CREDO. MooseFS is a distributed filesystem of high capacity.[1] It has built-in features of high reliability data storage: automatic creation of several low-level file replicas, regular monitoring of the data correctness (using internal checksums of low-level data chunks), and automatic re-creation of damaged replicas.

---

[1]More than 2 PB has been physically built for the CREDO project.

SZPAK extensions dedicated to the CREDO project include:

- division of the filesystem volume to so-called areas,
- possibility to allocate a particular file or directory in a recommended area and to request a specific number of low-level replicas for each file,
- power saving mechanisms, which allow powering off and on each of the areas,
- automatic calculation of file-level digital digest (currently SHA-256) for each file, and storing its value as an extended file attribute,
- a data carriers inventory (implemented with MySQL), which stores information on disks installed in SZPAK filesystem and collects S.M.A.R.T. messages from the disks,
- magnetic renewal of the data carrier, which consists in low-level rewriting of the data on the carrier,
- logical removal of a damaged or suspicious data carrier, with automatic low-level relocation of the data to other media,
- support for tape memory (LTO tape library).

MooseFS/SZPAK and Exchangeability of Technology

Tough in the current CREDO version dedicated filesystem SZPAK is used, the archive can work with other filesystems. SZPAK offers many dedicated extensions, but there are several methods to implement necessary functions using other filesystem:

- compliance with POSIX/SUS standard—if missing—can be achieved by use of additional layer, e.g. implemented in FUSE technology [11];
- other missing features can also be implemented in such additional layer;
- some missing features of the filesystem can be substituted by additional functions of specialized Persistence Management Subsystem (see Sect. 5), adapted to given filesystem;
- some of the missing operations (e.g. calculations of digital digests) may be executed by the Archive Management Subsystem (see Sect. 5), which is able to differentiate the course of executed operations depending on the used filesystem.

3.6 Archive Management Subsystem Technology

Archive management subsystems in CREDO are implemented on the basis of database management system (DBMS) Oracle.[2] Configuration and management data, event logs, and metadata of archival resources are stored in the database. The logic of the system is mostly coded in PL/SQL language, and executed inside the database. Such solution simplifies the system architecture, as additional layers, application technologies and means of inter-layer communication are unnecessary.

There are some exceptions, however. Adapter in file management subsystem (see Sect. 4.3) is implemented in Java (JEE) technology and runs on Glassfish application

---

[2]In the current CREDO version Oracle 12c Standard Edition One is used.

server. The adapter uses RESTful Web Services and XML messages to communicate with CREDO management subsystems. An optimization module (scheduler) is written in C++, since the performance of the calculations is important in this case.

Oracle and Exchangeability of Technology

CREDO implies the exchangeability of technology, but this can be reasonably accomplished only by the exchange of entire modules (subsystems or layers). Replacement of the DBMS results in at least need to create a new archive management subsystem. The creation of software portable between different DBMS has no technical justification and is not realistic with the reasonable cost. Need to replace the Oracle database to another DBMS in the foreseeable future is very unlikely, taking into account the excellent market share of the supplier. Such an exchange is possible at reasonable risk and cost, as the data structure is designed as a purely relational structure without use of the Oracle-specific extensions. This does not concern XML processing, where specific Oracle solutions are widely used, but there is practically no standardization of such solutions used in DBMS, so no standard approach can be used.

User Interface Technology

Oracle Application Express (ApEx) [12] tool is used to create a Web interface for end-users and for archive administrators. Oracle ApEx is a light declarative technology to create Web user interfaces to Oracle-based systems. It is built into the Oracle DBMS. As some kind of so-called 4-th generation application development tool, it provides high productivity and ease of application development and maintenance. The tool is constantly evolving, and seems to be more long-lasting than a variety of more or less popular frameworks, which quickly come and go.

The ApEx technology is, however, less known than other technologies for creation of end-user interfaces. It is also not typically deployed in information system layers: the ApEx engine runs within Oracle DBMS as a set of PL/SQL programs, and the WWW applications are made available to users via http(s) protocol, through proxy program Oracle REST Data Services (ORDS), which runs on some JEE–capable application server (e.g. Glassfish or Tomcat). The technology is also, of course, inseparably linked with Oracle databases.

ApEx and Exchangeability of Technology

Applications built with Oracle ApEx can be replaced by ones created with other technologies. The ability is assumed to create specialized applications for specific customers with technologies accepted by the customers. It is possible, because ApEx is used only for Web interface, and all the logic of archive operation is programmed outside ApEx in PL/SQL (in the database) and Java (on the filesystem side). This logic is constructed so that it can be easily used by applications written in any other technology, which can use standard interfaces to databases.

4   Architecture of the CREDO Long-Term Archive

The CREDO archive architecture is shown on Fig. 1.

4.1   Subsystems

The CREDO system is divided into clearly separated, loosely coupled subsystems, with well-defined tasks and clearly defined interfaces. Subsystems can internally have a layered structure.

Subsystem services are called locally through the exposed application programming interface (API), or remotely through RESTful Web services with XML messages.

This architecture promotes expandability and exchangeability of system components (in particular filesystems).

4.2   Data Flow in the CREDO Archive

In the CREDO long-term archive, the digital resources are transferred between the user environment and the archival storage through the buffer. This provides a higher level of security, as users have never any direct access to archival filesystems. It also enables efficient energy management in accordance with the idea of a deep archive (see chapter "The CREDO Project"), as only the archive buffer must be constantly powered on, and all the archival storage may be usually off-line. The data transfers between the buffer and the archival storage are done internally by the archive, and are optimized to minimize time windows the archival storage areas must be powered.

The filesystem provides basic means to ensure reliable storage of the data, such as low-level replication and low-level data consistency checks. The CREDO archive adds own high-level mechanisms: high-level replication/dislocation in multiple filesystems, digital digests management, magnetic refreshment of the data carriers, metadata management, data formats management, etc.

An end user (EU) uses standard file protocols to transfer files from his computer to dedicated proxy server EUOS (End User Operating System). EUOS system authenticates the connection and authorizes access to relevant parts of its filesystem. It does not provide any direct access to the filesystems of the archive.

Via an appropriate proxy protocol, a user can access his dedicated directory in the archive buffer from the EUOS system. Due to this solution, a significant level of safety of the connection is achieved, resulting from the architecture itself. A user has a direct access only to the dedicated EUOS server, located in the demilitarized zone, and the connection between this server and the internal CREDO filesystem is carried out solely by the proxy protocol, which is completely transparent to the user.

Fig. 1 Architecture of the CREDO archive

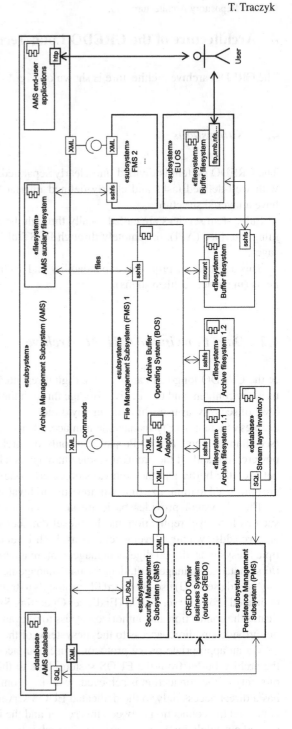

Data transfer between end-user programs and the CREDO buffer—via the EOUS and the proxy protocol—is done using common file protocols, e.g. scp or even ftp (it is secure, because the connection is possible only in VPN). Transfer between the buffer and the archival storage is managed by the archive, without any user intervention. Archival storage is made available to the archive management subsystems by mounting the appropriate directories of archival filesystems to the BOS (Buffer Operating System). It allows data transfer to/from remote archival filesystems and creating of dislocated high-level replicas in distinct distant locations.

4.3 Communication Between Subsystems

Communication interfaces between subsystems and layers must be precisely specified. Communication protocols should be designed so that they are as little depend on the implementation of subsystems and layers as possible. The protocols must operate in a distributed environment, and they should be open and permanent.

Such features can be provided by communication protocols based on the exchange of XML messages. As XML is a formalized language, which allows for precise control of syntax and data types, and is self-descriptive (contains metadata in the form of tag/attribute names), it allows to build communication protocols, which can be easy to understand (and possibly re-implement) also after a very long time.

One of the recognized standards of communication should be used to transfer XML messages. The CREDO system uses RESTful Web services. This solution can, however, be relatively easily replaced with any other means suitable for the transmission of text messages.

Subsystems should communicate using clearly defined channels. In CREDO, the communication between filesystems and other parts of the system is implemented using separated module, called 'adapter'. The adapter works in BOS system and performs necessary services on the buffer and the archival storage. It uses RESTful Web services and XML messages to communicate with other CREDO subsystems.

In future versions of CREDO, a cooperation of many (perhaps all) of the subsystems should be based on the idea of adapters. In the current version, this idea is implemented only in the case of interaction with filesystems, as these modules are potentially most vulnerable to exchange, and may be external products, independent of the CREDO project.

Interfaces between internal CREDO subsystems are constructed in a simplified, but technically rational manner. As these internal subsystems are currently implemented on the Oracle database platform, the communication bases on calls to APIs in PL/SQL, or on access to specified data using SQL. However, also in this case the interfaces are well-defined, and loose coupling between the subsystems is preserved. In the future it is possible, if necessary, to replace current methods of interaction by exchange of XML messages between appropriate adapters. Due to loose couplings and well-defined interfaces, such reimplementation of communication methods should not be costly or risky.

4.4 Deployment of CREDO Components

Deployment of CREDO components is shown in Fig. 2. The CREDO software operates on several servers, which should be located in at least two distant physical
locations.

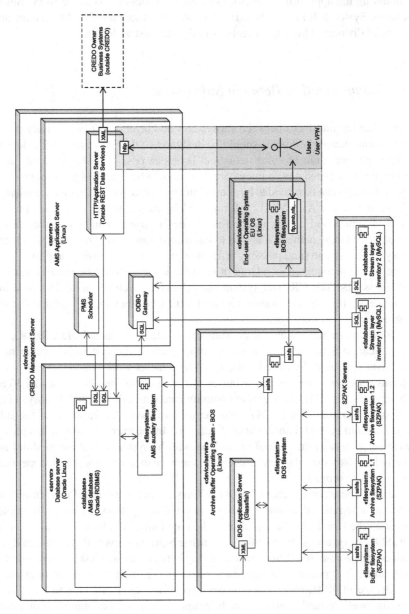

Fig. 2 Deployment of CREDO components

DBMS Oracle runs on the database server. The Oracle PL/SQL engine performs logic of the Archive Management Subsystem (AMS), the Persistence Management Subsystem (PMS), and partially of the Security Management Subsystem (SMS). Oracle Application Express (ApEx) tool, used to implement user interface, also runs inside Oracle DBMS. There is also an auxiliary filesystem that allows file-based data exchange between archive management subsystems and file management subsystems.

On the AMS application server Oracle REST Data Services (ORDS) runs, which provides access to APEX applications through HTTP protocol. ODBC gateway software allows for downloading data from the MySQL database of the SZPAK filesystem, which collects S.M.A.R.T. data from disks. The scheduler module also runs on the server, implementing optimization functions.

Buffer Operating System (BOS) runs on a separate machine. On its Glassfish application server, REST applications are deployed, which implement the functions of the adapter mentioned in Sect. 4.3. A filesystem of the BOS server acts only as a mountpoint for external filesystems: the buffer filesystem and archival storage filesystems.

The End User Operating System (EUOS), which is available to end users, runs on a separate server, located in a demilitarized zone. End users have restricted access to files in the buffer via EUOS and proxy protocol, and to CREDO Web applications through http(s) protocol. The access is possible only within virtual private network (VPN) dedicated to CREDO users.

The software of the filesystems runs on separate servers, in architecture appropriate for the MooseFS/SZPAK filesystem [10]. The buffer filesystem and at least one archival filesystem is mounted to the BOS. It is possible and recommended to mount two or more archival filesystems (possibly placed in remote locations), implemented as SZPAK instances or (in the future) in other technologies.

5 CREDO Subsystems

The basic principle of the CREDO architecture is the division of the system into well-separated subsystems. They must be as little dependent on each other as possible, have well-defined functions, well-defined and possibly invariable interfaces, and be composed of as simple modules as possible.

5.1 Archive Management Subsystem (AMS)

Archive Management Subsystem (AMS) is the main subsystem of a long-term repository. It performs basic functions of the archive in accordance with OAIS [1] recommendation. The subsystem interacts with other subsystems, managing their operation and using their services, e.g. scheduling implemented in PMS, file permissions management provided by SMS, etc.

The adapter in the BOS system (see Sect. 4.3) allows AMS to remotely perform operations on the filesystems in a manner appropriate for each specific filesystem. It communicates with AMS with simple XML protocol. The protocol is not dependent of used filesystem technology, so it is relatively easy to create a new adapter or replace the existing one. The adapter executes filesystem administrative commands (e.g. creates folders), and performs additional functions, which are not provided by the filesystem (e.g. calculates checksums, reads metadata from files, etc.). It also sends requests to filesystems, related to allocation, durability and security of files (e.g. requests an appropriate number of low-level replicas).

5.2 Security Management Subsystem (SMS)

Security Management Subsystem (SMS) interacts with AMS, fulfilling functions that require special privileges, e.g. knowledge of passwords. It is separated for safety reasons. The subsystem can be easily replaced with a version implemented with use of different technologies (e.g. based on LDAP) or with a component separated topologically (e.g. running on a specially protected server).

All CREDO subsystems operate in the internal trusted area. VPN provides encryption, hosts authentication etc. The adapter works in the filesystems with the least possible privileges. Security of access to archival resources is supervised by the SMS.

Outside of any archival session, no external user has any access to CREDO system. A user can communicate with the archive only in the context of explicitly commenced archival sessions, as recommended by OAIS standard. No external user ever achieve a direct access to CREDO filesystems, any access is possible only through the EUOS server and proxy protocol. SMS, in cooperation with AMS, establishes necessary privileges to appropriate directories for the user, only for the period of the archival session.

5.3 Persistence Management Subsystem (PMS)

Persistence Management Subsystem (PMS) prepares guidelines for AMS and filesystems on the reliability of the storage media, access schedule and energy efficiency. It takes decisions on the location of resource replicas, on data relocation, on replacement of unreliable data carriers, on archive operation schedule and power management.

PMS accepts orders from AMS containing requirements for archival packages storage durability. PMS creates then guidelines for the filesystem on allocation of package replicas in storage areas.

To effectively manage energy, PMS schedules access to the archive storage and sends commands to filesystems to turn the power on and off.

PMS collects data from filesystems to monitor the reliability of data carriers and evaluate their reliability. Then it manages relocation of the stored data from risky data carriers to more reliable ones, and need of data carriers exchange. PMS also manages periodical magnetic renewal of the data carriers to prevent their magnetic remanence degradation.

The current PMS version is dedicated to disk-based storage. It is, however, possible and quite easy to extend or exchange it by a new one prepared for a new storage hardware.

5.4 Business Subsystems

Business subsystems of the owner of the CREDO archive perform financial settlements, support formal and legal agreements, etc. They also store customers' and users' personal data, etc. They can also provide other business functions related to the operation of the archive.

These purely business functions or more complex workflows are not implemented as parts of the CREDO system. They can, however, interact with CREDO using dedicated interfaces, based on XML messages or on querying dedicated views in the CREDO database.

5.5 File Management Subsystem (FMS)

File Management Subsystem (FMS) consists of archival storage filesystems (stream layer), and auxiliary components: buffer filesystem, buffer operating system (BOS) with management system adapter (see Sect. 4.3) and inventory, containing information on data carriers and their reliability.

It is possible to have several File Management Subsystems in one CREDO archive. Each of FMSs may contain many archival filesystems, which should be deployed in more than one location and may be implemented in various technologies.

Archive Buffer

Access to the archive is possible only via buffer. Buffer and archive storage are separated, they should be placed in physically distinct filesystems. Independent SZPAK instance or other POSIX-compliant filesystem may be used as a buffer filesystem. The buffer works under the control of a dedicated operating system: Buffer Operating System (BOS).[3]

Software agent (AMS adapter) runs on an application server (Glassfish) as set of JEE applications. This agent operates with privileges of the buffer owner. It, however, operates on the archival storage with the lowest possible privileges. Operation of

---

[3]In the current version of CREDO Linux system is used.

automated processes on the archival storage with root privileges is not allowed at all; operation with root privileges on the buffer is strictly restricted.

The buffer filesystem is permanently mounted to the BOS. The archive storage directories are also mounted to the BOS, but with permissions that, in the initial state, prevent operations on the archival files. These restrictions are relaxed by SMS only for a period of an archival session, and only such permissions are granted, which are necessary to perform the actions of the session.

The end user has indirect restricted access (via proxy) only to dedicated directories in buffer storage, not to the archival storage.

6 Conclusion

The CREDO system architecture, consisted of loosely coupled subsystems with well-defined interfaces, and easily re-implementable modules and communication methods, promotes constant system development and adaptation to new technologies and new needs.

Thanks to careful separation of permanent architectural components from the ones which are subject to change, the system can be modernized fairly easily and safely, and the evolution of the archive can run smoothly.

References

1. Consultative Committee for Space Data Systems: Reference model for an open archival information system (OAIS). Recommended practice. (2012). http://public.ccsds.org/pubs/650x0m2.pdf. Access: 2016-10-25.
2. World Wide Web Consortium (W3C): Extensible markup language (XML) 1.0 (fifth edition) (2008). http://www.w3.org/TR/REC-xml. Access: 2016-10-25.
3. World Wide Web Consortium (W3C): XML Schema version 1.1 (2012). http://www.w3.org/XML/Schema. Access: 2016-10-25.
4. World Wide Web Consortium (W3C): Namespaces in XML 1.0 (third edition) (2009). http://www.w3.org/TR/REC-xml-names. Access: 2016-10-25.
5. World Wide Web Consortium (W3C): XQuery 1.0: An XML query language (second edition) (2010). http://www.w3.org/TR/xquery/. Access: 2016-10-25.
6. Library of Congress: Metadata encoding & transmission standard. http://www.loc.gov/standards/mets. Access: 2016-10-25.
7. PREMIS Editorial Committee: PREMIS data dictionary for preservation metadata (2012). http://www.loc.gov/standards/premis/v2/premis-2-2.pdf. Access: 2016-10-25.
8. The Open Group Base Specifications Issue 7 (2013). http://pubs.opengroup.org/onlinepubs/9699919799. Access: 2016-10-25.
9. Skytechnology Ltd. (2016). http://skytechnology.pl. Access: 2016-10-25.
10. About MooseFS. http://www.moosefs.org. Access: 2016-10-25.
11. Filesystem in userspace (FUSE). http://github.com/libfuse/libfuse. Access: 2016-10-25.
12. Oracle Corp.: Oracle application express. http://apex.oracle.com. Access: 2016-10-31.

Information Processing in CREDO Long-Term Archive

Tomasz Traczyk

Abstract CREDO processes archived information in several types of archival and administrative sessions. This chapter describes these sessions, and logical as well as physical structures that are used to store and process archived resources.

1 Introduction

The CREDO archive has to process delivered and archived information to provide the basic functions of the archive as well as to ensure data longevity, storage reliability, energy efficiency, etc.

The information is processed in sessions, and the archived data are organized in archival packages, as required by OAIS [1] recommendation.

2 Structure of the CREDO Archive

2.1 Physical Structure of the CREDO Archive

CREDO archival storage can consist of many filesystems (see chapter "CREDO Repository Architecture"). Each filesystem is divided into so-called storage areas. The area is the smallest part of the archive, which can be independently switched on or off. The area consists usually of many data carriers (see Fig. 1). Each area should be reasonably uniform in terms of data carriers performance and reliability.

The data can be stored in many low-level copies, managed internally by the filesystem and transparent to users.

T. Traczyk (✉)
Institute of Control and Computation Engineering,
Warsaw University of Technology, Warsaw, Poland
e-mail: T.Traczyk@ia.pw.edu.pl

© Springer International Publishing AG 2017   93
T. Traczyk et al. (eds.), *Digital Preservation: Putting It to Work*,
Studies in Computational Intelligence 700, DOI 10.1007/978-3-319-51801-5_5

Fig. 1 Relationships
between data carriers,
storage areas and regions,
and package replicas

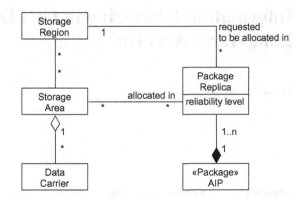

2.2 Logical Structure of the CREDO Archive

2.2.1 Resources in the Archive

Digital resources in the CREDO archive are organized in form of archival packages, conforming to OAIS recommendations [1].

SIP

A Submission Information Package (SIP) is delivered to the archive by a so-called producer. The archive checks package completeness and consistency of delivered data files with their delivered metadata. Then the package is supplemented with additional metadata generated by CREDO: a manifest file, defining package contents and structure (in current version of CREDO in METS format) and preservation metadata (in form of PREMIS file). The data flow between archival packages is shown in Fig. 2.

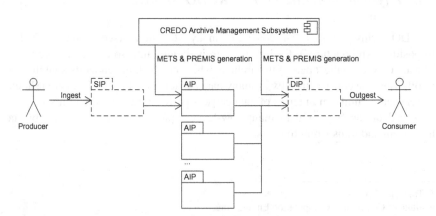

Fig. 2 Data flow (simplified) between CREDO archival packages

SIP packages are temporary, they exist only during Ingest sessions.

AIP

The checked and completed package is copied to the archival storage as an Archival Information Package (AIP). The package consists of data and metadata files originally delivered by the resource producer, and of the package manifest file and preservation metadata file. The entire contents of the SIP package and the original structure of subdirectories is preserved; the additional files are added to the root directory of the AIP package.

In CREDO the AIP package is represented simply by a branch in archival storage directory tree. It is not compressed or packed in any other way, so that the package form does not depend on any specific file format. It supports long-term usefulness of the data and simplifies (or even enables) operation on very large files (e.g. of many terabytes).

AIP packages are permanently stored in the archive.

DIP

When archived resources are to be delivered to a consumer, they are copied together with their metadata to a Dissemination Information Package (DIP). A DIP package can include contents of many AIP packages, which are copied to distinct DIP subdirectories. The DIP package is also supplemented with necessary additional metadata generated by CREDO: a manifest file and a preservation metadata file. Then the package is made available for the consumer to download.

DIP packages are temporary, they exist only during Outgest sessions.

2.2.2 Fonds and Owners

Each archival package has exactly one owner—one of the archive customers (see Fig. 3). Usually it is the customer, who delivered the package to the archive.

Customers' packages are organized in so-called fonds: each package belongs to one fond. Each fond belongs to one customer, but a customer can have many fonds. It is recommended that packages related in content were stored in one fond.

In CREDO each fond is stored in separate folders, one in each filesystem used to preserve the fond. AIP packages belonging to the fond are stored in subdirectories of these folders. Furthermore, the relationship between the fond and its packages and files is also stored in the archive database and in extended attributes of the files. Therefore, even if a file is accidentally moved, it can be found and placed in the right place.

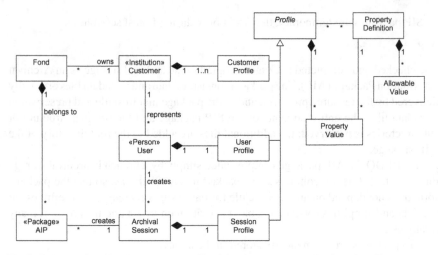

Fig. 3 Relationships (simplified) between CREDO customers, users, fonds and profiles

2.2.3 Customers, Users and Profiles

A customer is an institution or organization, which in CREDO plays role of a producer or a consumer of archived resources (by OAIS [1] terminology). A user is an account of a person acting in CREDO for a specific single customer (see Fig. 3).

CREDO only stores the identification data for customers and users. Any other details (including personal data) can be stored in external systems, e.g. CREDO owner business systems.

The customer is identified by an alphanumeric code assigned by the administrator, a customer name acts only as an auxiliary descriptor. The user has an artificial numeric identifier and a login name, which is only used to authenticate him. Auxiliary external identifiers relates customers and users to entities defined in external systems.

Customers' and Users' Profiles

Customer profile defines a set of properties, which control sessions committed due to a given customer. Each customer can have several profiles, e.g. defining different storage requirements for various classes of resources. A user representing the customer can use customer's profiles. A property value set in customer profile can—if it is allowed by property definition—be overwritten in user's profile.

Profiles allow customers and users to define their standard requirements for level of reliability, storage regions, allowed file formats, metadata processing, etc.

Session Properties

Archival sessions work in the context of specific profiles. When a session is created, one of the customer's profiles must be selected. Values of session properties are inherited from selected customer profile and current user profile to the session profile. Values of some properties can also be set or changed on per-session basis.

The resultant values of the properties in session profile control operation of the session.

2.3 Logical Structure of the Archival Storage

Storage areas (see Sect. 2.1) are grouped into storage regions, which contain areas having similar features (e.g. offering similar security or performance, located in the same data center, dedicated to the same customer, etc.).

The user may request (within granted privileges) regions, in which the package is to be stored. Archival packages are stored in several high level replicas. Requests of the region apply to each of the replicas separately (see Fig. 1), allowing precise control of replicas dislocation, diversification of technologies, use of dedicated filesystems, etc. Storage regions are virtual entities, defined only to be used in such requests, in contrast to the storage areas, which are physical structures.

When saving the AIP package, the CREDO archive requests allocation on specific storage areas, one for each package replica. The area is selected from all areas belonging to the requested region by optimization algorithm, so as to comply with the region request, to minimize energy consumption and to store the data on the most reliable media available within given constraints.

The filesystem tries to save the data in accordance with this area allocation request. In some cases, e.g. if the requested allocation is impossible or after some relocations forced by hardware failures, a package can be actually allocated in other area or even in more than one area.

The filesystem also adjusts the number of low-level copies (see Sect. 2.1) to the reliability level requested for each replica.

3 Open Model of Archival Sessions

Archival data processing is done only in sessions, according to OAIS recommendation [1].

The archive performs sessions of several types. These types and appropriate actions are defined in the archive database, in flexible data structures, so that it is quite easy to modify the algorithm of the session and it is relatively easy to create new types of sessions.

3.1 Session Types

In the CREDO archive two groups of sessions exist: archival sessions, which create, search or transform archival packages, and administrative sessions that perform various auxiliary activities.

Archival Sessions

Archival sessions are of the following types.

- Ingest sessions—receive SIP packages delivered by producers, check and complement the packages, transform them into AIP packages, preserve them in the archive and store selected metadata in the archive database.
- Search sessions—search AIP packages using metadata stored in the archive database.
- Outgest sessions—get requested resources from AIP packages, copy them to DIP packages, supplement them with necessary metadata and make available to consumers.

In future versions of CREDO other types of archival sessions may be added, e.g. for format transformations.

Archival sessions are performed by archive users, who use Web-based end-user application to create and control the sessions.

Administrative Sessions

Administrative sessions in CREDO include:

- package state check sessions—check completeness of the stored AIP replicas and correctness of the digital digests of the archived files;
- magnetic renewal sessions—periodically request magnetic renewal of data carriers;
- reliability check sessions—analyze data collected from data carriers (e.g. S.M.A.-R.T. data from disks), evaluate reliability of data carriers and storage areas;
- area migration sessions—work out recommendations for archival data relocation to more reliable areas.

These sessions are performed periodically by archive administrators.

3.2 Logical and Physical Sessions

Sessions (in the meaning of the OAIS standard) can take a very long time, even several days if a large volume of archival data is to be processed. We call these sessions 'logical sessions', as they perform logic of specific archive operations and maintain session state.

It would be highly impractical if the user had to be connected to the system throughout the duration of such session. Therefore, a logical session should be able to perform despite user logins and logouts. Periods when a user is connected to the system and to one of his logical sessions we call 'physical sessions'.

Any CREDO logical session may last during many physical sessions and periods between them. Of course, the state of the logical session is preserved throughout the session, and the operations of the session are performed (whenever possible) also when the user is disconnected from the system. When the user logs back to the system (i.e. starts new physical session), he can select one of his active logical sessions. Each physical session is in fact associated with exactly one logical session.

3.2.1 Actions—Components of Logical Session

Each logical session performs one main procedure, suitable for the type of session, which consists of some subordinate procedures. Each subordinate procedure consists of several operations, and each operation may contain several operation steps. All these activities are called actions. Figure 4 shows action hierarchy in CREDO.

Actions are defined in flexible data structure, which allows to specify:

- sequence of the actions within parent action, i.e. procedures within main procedure, operations within procedures, etc.;
- dependencies that must be satisfied to start and finish the action, defined like dependencies in precedence diagrams [3], i.e. start-to-start, finish-to-start, start-to-finish and finish-to-finish;
- minimum and maximum expected time of execution;
- questions displayed to the user together with answers to choose;
- conditions (function calls) to be executed before the action starts and before it finishes, e.g. checking user answer to asked question or some result of preceding action;
- program (procedure call) to be executed by the action;
- etc.

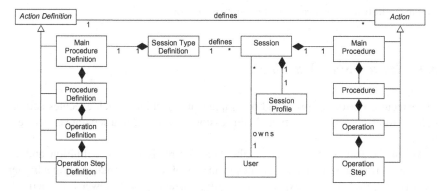

Fig. 4 Actions in CREDO

In the current version of CREDO procedures and functions called by the actions are coded in PL/SQL language, but in future versions calls to subroutines written in other languages may be added.

3.3 Session Management

CREDO sessions are managed by an executor, which is a module of Archive Management Subsystem (AMS, see chapter "CREDO Repository Architecture"). The executor co-operates with another AMS module: a scheduler (described in chapter "Power Efficiency and Scheduling Access to the Archive"), which optimizes access to the archival storage.

3.3.1 Limited Process Model

It can be noticed that the actions make up some structure similar to project network, precedence diagram or flow chart. The network of actions is, however, more restricted than in these popular process models, because it has to be adapted to the capabilities of scheduling algorithms used to optimize power consumption and storage allocation. Too general process model leads to unreasonably complex and inefficient algorithms, so the generality of the model has to be limited. In CREDO the action network cannot contain any branches. There is, however, a possibility to conditionally skip any action, and to conditionally restart or abort a procedure. These capabilities have proven to be sufficient to model the necessary session logic, along with appropriate responses to user actions.

The optimization algorithm controls the operation of the archive by scheduling its actions on operation level. The operation—if defined as being scheduled—is executed in time proposed by the scheduler (or later, if it must wait for start conditions to be satisfied). Actions of other levels are always executed as soon as possible (taking into account the action sequence and start conditions).

3.4 Session State Machine

The executor performs some kind of simple state machine, which executes actions defined for the given session type. Each session is executed in a separate concurrent thread.

When the user starts a session, the executor reads session definition and its assigned action definitions, and creates instances of actions. When starting a higher level action, the executor expands it, creating instances of its lower level actions. When an operation is to be executed on multiple objects, e.g. on several replicas of a package, the executor multiplies instances of the operation, and assigns each of the copies to one of the objects. Such multiplied operations are independently scheduled, but they may be executed in parallel.

After expanding the action, the executor tries to execute subsequent subordinate actions, appropriately checking their start dependencies and conditions. When all the subordinate actions are finished, and own logic (procedure) of the action is completed, finish dependencies and conditions are checked, and—if satisfied—the action finishes. If start or finish dependencies or conditions are not satisfied, an execution of the action is suspended and the executor tries to perform other actions, e.g. parallel ones.

If an operation is defined as being scheduled, the executor waits for the scheduler to work out the proposed start time for the operation. Next, when a scheduled operation is in order to start, the executor waits for a time scheduled for this action.

3.4.1   Interaction with Users

During the execution of the session, its state is recorded to the database. End-user applications should read the state and present appropriate forms, which allow users to perform necessary actions, e.g. examine or enter necessary data.

Sometimes a user decision is needed, which may change an execution path, e.g. may cause some procedure to be skipped or restarted. In such cases, a question text, together with possible answers, can be read from the definition of current action. The application should display the question and store the answer to the database, in current action record. The executor acts accordingly to the answer.

During selected phases of a session, the user is able to abort the session. The administrator can hold or abort the session in any state, and in most cases he can also resume the session from the point of suspension.

3.4.2   Exception Handling

In the CREDO system, a very conservative model of exception handling is applied. There is no automatic or semi-automatic response to emergency situations. As the archive is obliged to ensure the highest level of data security and accuracy, each problem has to be examined and appropriate decisions have to be taken by a human expert.

If an uncritical error occurs, the execution of the session is suspended. Appropriate messages are shown to the user and written to the journal tables. The archive administrator can then examine the session state, trail and logs, possibly solve the encountered problems, and resume the session from the point of suspension, or decide to abort it.

If a critical error occurs, the session is automatically aborted. It is however possible, by intervention of the administrator, to force resume of the session or to start a new session, which uses the data provided to the killed one.

3.4.3   Execution Trail and Logs

The execution trail, including data of actions and their state, is immediately and permanently written to the archive database, so it is possible to preview all the activities of the system, both on-line and later.

Events, which happen during the execution of the session, are logged into journal tables. Several logging levels are available, from fatal errors only logging to very detailed debug logging.

The journal tables contain also detailed information on all REST calls to the AMS adapter in BOS (see chapter "CREDO Repository Architecture"), together with their results and complete XML messages exchanged between AMS and the adapter. It enables the administrators to track details of system work, e.g. in case of some problems or errors.

4   Archival Sessions in CREDO

In the following sections typical scenarios of the three types of archival sessions are presented. 'Typical' means that the session runs without any problems and all steps, which require user approval, are confirmed. The scenarios, shown as UML sequence diagrams, are simplified to make the diagrams more clear, e.g. most feedback messages and requests to the database are omitted.

Archival session in the CREDO archive is always launched by a user. The session is executed on behalf of a CREDO customer the user is representative of.

4.1   Ingest Session

A scenario of typical Ingest session is shown in Fig. 5.

Before the session starts, the user has to provide several session parameters, which control its execution, e.g. requirements for level of reliability, storage regions, allowed file formats, metadata processing, etc. Initial values of the parameters are copied from customer's and user's profiles, then the settings may be adjusted to the needs of the current session.

When the session starts, a folder for new SIP package is created in the buffer, and the user is allowed to copy (safely: in VPN, via proxy server and proxy protocol) the delivered SIP content to the folder.

The metadata of the package can then be read from metadata files delivered in SIP package, extracted from data files (from so-called embedded metadata), and entered or corrected manually using end-user Web-based application.

The actual content of the delivered SIP package is checked against what is declared in the metadata. File formats and digital digests are compared. If everything is fine, CREDO creates its own manifest file in METS format and preservation metadata

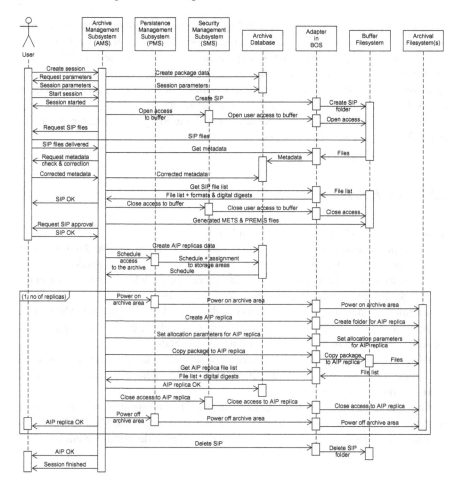

Fig. 5 Typical Ingest session scenario (simplified)

file in PREMIS format, and adds them to the SIP package. The user can examine the content of the package and confirm its correctness.

If something goes wrong, the user can restart the SIP delivery procedure and complete or change content of the SIP package or its metadata.

Next the AIP creation procedure starts, which may take much time. This part of the session does not, fortunately, require the constant attention of the user.

First, the schedule is calculated, which ensures energy efficiency and use of as reliable data carriers as possible. The schedule contains therefore not only the timetable of necessary operations, but also recommendations for usage of storage areas.

Since the AIP package should usually be stored in many replicas (possibly in more than one filesystem), next part of the session algorithm must be multiplied. The section shown as an interaction group in Fig. 5 is executed as many times as replicas are requested. The operations on different replicas may be executed in parallel.

For each replica, the necessary storage area is powered on the schedule. After the files are copied to the destination area, copy correctness is checked. Next access rights on files and folders are changed, so the unprivileged user or program cannot access the data. Finally, the storage area is powered off.

When all AIP replicas are successfully created, the SIP package becomes useless and is deleted. The session is finished.

4.2 Search Session

A scenario of Search session is shown in Fig. 6.

This type of session uses only the metadata stored in the archive database. The archival storage is not affected, therefore no access optimization is needed.

The search procedure can be repeated several times, when the user can enter or change the query, using easy to use Web-based interface. Each user query is transformed to an appropriate SQL query and executed in the database. The results are presented to the user. When the results are satisfactory, the last user query and the results are saved in the database and the session ends.

The query entered by the user can consist of several conditions, concerning basic properties of the resources (e.g. identifiers, ownership, filenames, date of creation), as well as their metadata.

Any simple metadata element defined in the CREDO database can be used in the search. In particular, simple Dublin Core metadata terms (DCMES [2]) are predefined and are recommended to be used when searching descriptive metadata.

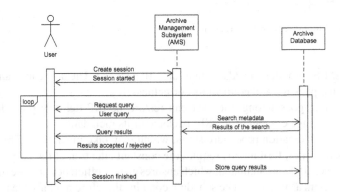

Fig. 6 Search session scenario (simplified)

Complex metadata stored in form of XML documents can also be searched using conditions written in XQuery language.

The stored query can be used as a start point in subsequent Search sessions. The stored search results can be used to specify requested resources in later Outgest sessions.

4.3   Outgest Session

A scenario of typical Outgest session is shown in Fig. 7.

When the session starts, its parameters must be set, which include the list of requested resources (in current version of CREDO it is a list of AIP packages). Results of previous Search sessions can easily be added to the list.

Next the availability of the requested resources is checked and the resources, which cannot be delivered to the user (due to insufficient privileges), are removed from the list. If the resultant list is not satisfactory, the session can be stopped or the procedure can be restarted, and the list of requested resources modified.

Next the DIP creation procedure starts, which may take much time. This part of the session does not, however, require the constant attention of the user.

First, the schedule is calculated to optimize energy consumption. The resultant schedule contains the timetable of operations, with decisions which replica of each necessary package should be used.

Since the DIP package usually consists of many AIPs, next part of the session algorithm must be multiplied. The section shown as an interaction group in Fig. 7 is executed as many times as AIPs are requested. The operations on different AIPs may be executed in parallel.

For each source AIP, the necessary storage area, which contain the replica to be read, is powered on the schedule. Access rights are changed so that the data can be read by copy process. After AIP files are copied to a DIP subdirectory in the buffer, copy correctness is checked. Next restrictive access rights on files and folders are restored and the storage area is powered off.

Finally, CREDO generates DIP metadata files: package manifest in METS format and preservation metadata in PREMIS format.

The DIP package is ready, so the access rights to the buffer are changed to enable the user read the package (via proxy server and proxy protocol). The user downloads the package and announces the completion of the operation. The DIP package is now useless and is deleted. The session is finished.

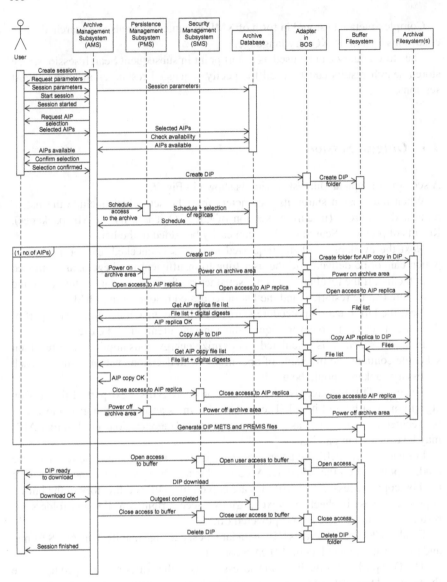

Fig. 7 Typical Outgest session scenario (simplified)

5 CREDO User Interface

The CREDO system is available for users through end-user applications, and administrative applications.

5.1 CREDO End-User Applications

By assumption, end-user applications should be created for particular CREDO customers. The applications may therefore be specifically adapted to customers' needs. These applications can be crated in various technologies, preferred by users. They co-operate with CREDO services by SQL CRUD[1] operations on appropriate tables in CREDO database, and calls to PL/SQL API subroutines. Therefore it is relatively easy to create such applications in any technology, which enables access to relational databases.

A generic end-user application has been however created, and can serve to present CREDO capabilities. It can also be used in initial period of CREDO utilization, to gain experience or before customized applications are ready. The application is Web-based and uses Oracle Application Express (ApEx) technology (see chapter "CREDO Repository Architecture").

CREDO end-user applications perform following tasks:

- management of Ingest, Search and Outgest sessions,
- preview of session history,
- preview of the results of user's search sessions.

Future end-user applications may also facilitate massive data transfers between users' environments and the CREDO buffer.

5.2 Administrative Applications

Administrative applications were developed with use of Oracle ApEx technology. They may be further developed, but there is not need to change the technology at least until the Oracle database is used.

Configuration Application

The CREDO system bases on flexible data structures and logic partially controlled by configuration data. Therefore, the system needs complex configuration data to be set.

Configuration application is used by CREDO developers and privileged administrators to configure the CREDO repository and to define session logic. It can be used to:

- define configuration parameters and set their values,
- define properties of users' and customers' profiles, sessions, storage objects, etc.,
- define storage regions,
- define session types, their actions and action sequences.

Administration Application

Administration application is used by CREDO administrators to perform their daily tasks:

---

[1]Create, read, update, delete.

- supervision over the operation of the system,
- preview event logs (errors, warnings, etc.),
- preview execution state of CREDO sessions,
- pausing, aborting and resuming sessions,
- launching administrative sessions,
- defining customers and users, and setting their profiles.

Persistence Management Subsystem Administration Application

This application is separated because of necessary separation and exchangeability of the PMS subsystem. PMS performs in fact these functions of CREDO, whose heavy dependence on managed filesystem and its features cannot be avoided.

PMS administration application is therefore used by CREDO administrators to perform filesystem-specific tasks:

- defining storage areas,
- preview and editing data of storage devices and their parameters,
- preview data usage (high-level replicas, storage areas) by sessions,
- preview allocation of archival packages.

6 Conclusion

CREDO information processing model fully conforms to requirements of OAIS recommendation.

Basic archive activities are performed in Ingest, Search and Outgest sessions. Many auxiliary tasks are provided by several types of administrative sessions.

Though main archive functions have been implemented, the system is flexible and can be easily adapted to new requirements. Due to open session model and flexible data structures, it is possible to extend the operations of the archive or to adjust them to specific needs. For particular archive customers customized end-user applications can be created, which can be easily interfaced to the CREDO management subsystems.

References

1. Consultative Committee for Space Data Systems: Reference model for an open archival information system (OAIS). Recommended practice. (2012). URL http://public.ccsds.org/pubs/650x0m2.pdf. Access: 2016-10-25.
2. Dublin Core Metadata Initiative: Dublin core metadata element set, version 1.1 (2012). URL http://dublincore.org/documents/dces. Access: 2016-10-25.
3. Wikipedia: Precedence diagram method. URL http://en.wikipedia.org/wiki/Precedence_diagram_method. Access: 2016-10-25.

Metadata in CREDO Long-Term Archive

Tomasz Traczyk and Grzegorz Płoszajski

Abstract In long-term archiving stored resources have to be accompanied by information which enables their searching, helps proper preservation and maintenance, and may be used to examine the correctness of the preservation. This information, called metadata, has to be therefore stored and processed in the archive. This chapter describes metadata preservation and processing in the CREDO archive. The archive stores metadata in archival packages. Since we are dealing with a deep archive, this type of storage is usually off-line, and any access to archived data is slow. To enable effective search and efficient maintenance of the archive, copies of selected metadata are stored also in on-line archive database.

1 Introduction

Storing and appropriate processing of metadata is crucial for proper digital archive operation, as stated in chapter "Metadata in Long-Term Digital Preservation".

Any digital archive must store metadata, needed to verify the completeness, authenticity and non-repudiation of resources collected in the archive.

Long-term archive is not a system strongly supporting a wide dissemination of stored resources. The primary objective is to preserve the resources over possibly long time, not to publish them. Search functions of the archive are limited and, by assumption, not intended for public use. Therefore, the search need not provide a high performance.

Searchable information must be however present in the archive, making it possible to find the needed information without detailed knowledge of the exact location of the wanted data. The searchability of the archival resources is particularly important

T. Traczyk (✉) · G. Płoszajski
Institute of Control and Computation Engineering, Warsaw University of Technology,
Warsaw, Poland
e-mail: T.Traczyk@ia.pw.edu.pl

G. Płoszajski
e-mail: G.Ploszajski@ia.pw.edu.pl

© Springer International Publishing AG 2017   109
T. Traczyk et al. (eds.), *Digital Preservation: Putting It to Work*,
Studies in Computational Intelligence 700, DOI 10.1007/978-3-319-51801-5_6

in a long-term archive, as the archived information should be available even after dozens of years, when the creator of the information no longer exists.

2 Metadata in the CREDO Archive

Since we are dealing with a deep archive, searchable information (i.e. metadata) preserved in the archived resources cannot be directly used, because archive files are stored in parts of archive storage, which most of the time are turned off to ensure energy efficiency. Therefore, the key metadata, allowing search and identification of resources, must be stored separately, in an on-line database.

Metadata can be assigned to each level of the hierarchy of objects in the archive: packages, resources and individual files.

2.1 Metadata Types in CREDO Long-Term Archive

2.1.1 Technical Metadata

The database contains metadata necessary for the proper operation of the archive, as resource identifiers, digital digests of archived files, file format declarations, etc.

Identifiers

Digital object (a package, resource or file) has one internal CREDO identifier, but it can have many various external identifiers, stored as pairs: identifier type (value chosen from the dictionary: URL, DOI, etc.), and value of the identifier.

File Format

For digital files, a format can be declared. The format is specified using MIME Type [7] of the file. On the basis of MIME type, the archive compares actual file format with declared one, and takes some decisions regarding the processing of the file, e.g. embedded metadata acquisition (see Sect. 2.2.2).

2.1.2 Preservation Metadata

All the operations executed during archival sessions are logged in a journal in the archive database. The content of the journal may be considered as internal preservation metadata.

This metadata are written to XML files (currently in PREMIS format [9]) and stored in AIP packages, then may be made available as part of DIP packages. Together with original preservation metadata delivered in SIP package, they constitute a complete history of resource preservation.

2.1.3 Descriptive Metadata

Though it is possible to define arbitrary metadata elements, descriptive metadata compatible with simple Dublin Core Metadata Element Set standard [6] are preferred in the CREDO archive, as they allow user to query the database in simple, well-known and uniform way. Therefore, descriptive metadata should preferably be delivered to the archive in simple Dublin Core standard.

Repeatable metadata elements, i.e. having multiple values, are allowed: selected metadata elements can have several occurrences assigned to the same archival object, but with different values.

2.1.4 System Metadata—Documentation of Data Formats

Some file formats are recommended for use in digital archives, some formats may be accepted, and other formats may be considered unacceptable. In CREDO, recommended and accepted formats are defined at system level, and it may be overridden at customer level. It can also be set whether the other formats are unacceptable or tolerated.

For a file format defined in the CREDO archive, a link can be specified to the system-owned archival package, which contains documentation of this format. Such system packages are archived in dedicated system fonds, owned by the archive. Documentation concerning metadata standards and metadata file formats may also be preserved in this way.

In a distant future this documentation may allow or facilitate the read and correct interpretation of the content of the archived files, therefore may ensure their archiving at 'content preservation' level.

2.2 Sources of Metadata

2.2.1 Metadata Files in SIP

A delivered SIP (Submission Information Package [4]) can (and even should) contain metadata files. In CREDO following types of metadata files are expected:

- a manifest (currently in METS format [8]) for the SIP package, which declares the contents of the package;
- preservation metadata (in PREMIS format), which describes the origin and previous history of the delivered resources;
- and other metadata files in any XML format.

XML-based metadata files are preferred, as this format is non-proprietary, text-only, and self-descriptive, so it ensures a proper data read and interpretation after many years.

These metadata are stored in archival packages. Some of them can be analyzed by the archive (e.g. the manifest file), and partially stored in relational structures of the archive database. Some of them may also be stored in the database in form of XML documents.

2.2.2 Embedded Metadata in SIP

An archived file can contain so-called embedded metadata. Many popular file formats, e.g. PDF, graphic formats (JPEG, TIFF, etc.), audio and video formats (MP3, WAV, MPEG4, etc.) enable embedding technical and/or descriptive metadata in the file.

These metadata can be selectively extracted in form of XML document and stored as XML in the database. Selected elements of the embedded metadata can also be stored in the relational structure, e.g. mapped to standard Dublin Core elements.

3 Metadata Preservation in CREDO

In the CREDO archive metadata are stored in several ways.

3.1 Metadata in Archival Packages

The metadata of the resources may be delivered in SIP packages and stored in AIP packages, together with the described resources. In the CREDO archive all files delivered in SIP package, including metadata files, are preserved without any changes in AIP package.

The metadata are usually stored in separate files (preferably in XML format). The CREDO archive can interpret the contents of delivered metadata files in METS standard, and also creates its own metadata files in METS and PREMIS standards. The archival packages, however, can also contain metadata files in other formats. Metadata files are stored in the archival storage and can—under certain conditions— be copied to the archive database to become searchable.

Archived data files in some formats, e.g. graphic files, may also internally contain embedded metadata. Selected embedded metadata may be read by the archive and loaded into the archive database to enable searching.

Some selected metadata, related to resource ownership, identification, and belonging to specific AIP package, are additionally stored in so-called extended attributes of the archived files. This metadata are intended for the case of package damage, e.g. accidental transfer of the archived file from one package to another.

3.2   Metadata in CREDO Database

Metadata, which should be searchable or available on-line, are stored in the archive database.

3.3   Metadata in Relational Structures

The database stores metadata in relational structures, which allow search using standard SQL queries, and possibly full-text search. Metadata stored in this way can be very efficiently searched and reported, but their complexity is limited.

Basic technical metadata of archived files: paths and names, volumes, creation dates, ownership and belonging to appropriate fonds, format specifiers, digital digests, etc., are stored in specialized relational structure.

Preservation metadata are stored in archive journal tables, as mentioned above.

Also other metadata can be stored in relational structures, e.g. descriptive metadata. The data structure for this kind of metadata is very flexible. It is so-called generic structure, configured by dictionaries, which may be adjusted by system administrators, e.g. new metadata elements can be added when necessary. In the initial version of the CREDO archive, entries are defined for Dublin Core Metadata Element Set.

3.4   Metadata in XML Documents

Metadata in the archive database can also be stored in form of XML documents. Metadata stored in this manner may be very complex. They can be quite effectively searched using XQuery language, but such search is less effective than SQL queries to relational structures. In the case of deep archive, search speed is, however, not very significant.

The XML documents are connected to the data objects that describe files, resources or packages.

Metadata delivered as XML files can be copied into the CREDO database. In principle, such copies should be stored for supplied manifest documents, containing specifications of files in SIP packages (e.g. as METS files). User can also indicate other metadata files to be copied into the database, e.g. PREMIS files provided in SIP, which contain provenance metadata describing the origin of digital objects.

The database may also store XML documents containing embedded metadata extracted from archived files.

4   Metadata Processing in CREDO

A SIP package delivered by a user is stored in a temporary directory created for an Ingest session. The package usually contains metadata files and embedded metadata in other files. The metadata can also be supplied manually during the Ingest session (see next section).

Delivered metadata are stored in archival packages, selectively copied into the archive database, and—when needed—disseminated in Outgest sessions. Metadata flow in the archive is shown in Fig. 1.

4.1   Metadata Processing in Ingest Session

The list of delivered files, and the metadata related to them (and to the whole package) may be acquired—and this is the preferred method—from the delivered manifest file, as described below.

If the manifest is not delivered, the list of files and other metadata could be supplied manually, using Web based end-user application. The archive may, on demand, facilitate this process by automatic generation of file list, based on the actual content of the delivered SIP. The generated list contains also digital digests and file format specifiers obtained from the files. The user can then manually add missing metadata.

4.1.1   Metadata Files Processing

Manifest File Processing

It is expected that a delivered SIP package contains a manifest metadata file, which provides declarations of files in the package. Each declaration consists of the file name (with relative path), digital digest (SHA-256 is currently accepted), declared file format (as MIME type), and optionally descriptive metadata. Current version of CREDO accepts manifests in METS version 1.6 format [8], but in the future other formats may be quite easily added, e.g. XFDU [3].

This manifest file is downloaded into the database and analyzed. A list of declared files is created in the database, and—if possible—descriptive metadata of the package are loaded.

Preservation Metadata Processing

Digital resources delivered in the SIP package can have their initial preservation metadata, which describe their origin and former preservation activities. In CREDO, use of PREMIS format version 2.2 [9] is preferred for this purpose.

Delivered PREMIS file should contain the initial information about the digital object, useful for long-term storage. In particular, it should include the information

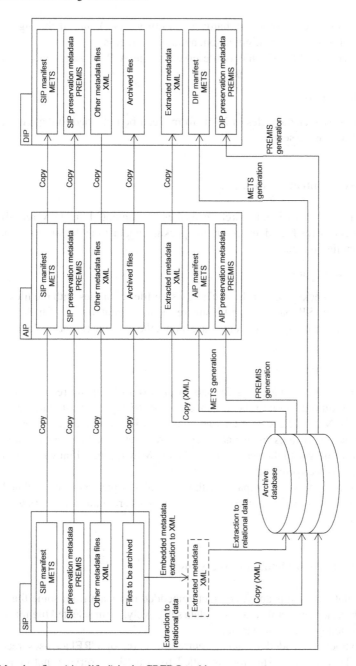

Fig. 1 Metadata flow (simplified) in the CREDO archive

about the object creation (date, hardware, software, etc.), and, if the digital object has been produced by the digitization, about this process.

This original PREMIS file is preserved in the AIP package, and can also be uploaded to the archive database.

Other Metadata Files

Selected other metadata files in XML format can be uploaded to the database (see Sect. 3.4) to enable search.

4.1.2 SIP Content Verification

During the Ingest session, the archive verifies the content of the delivered SIP package, and compares files delivered in the package against the declarations contained in the supplied metadata.

File List Verification

The archive checks whether the declared list of files complies with the files actually delivered in the SIP package, and compares actual digital digests of files against the values declared. If any discrepancies are found, the package (or the declaration) must be corrected.

File Format Verification

The CREDO system identifies formats of archived files using recognized DROID tool [5]. Acquired format-related metadata: MIME type, format name, and format PUID (PRONOM's Persistent Unique Identifier [5]) are stored in the database.

If formats of the files are declared in delivered metadata, the archive compares the declared formats with the acquired ones, using MIME types. If any inconsistency is found, it must be corrected.

File formats are also compared with the list of acceptable formats defined for the customer or for the entire archive. If inconsistencies are found, appropriate warnings are generated.

4.1.3 Embedded Metadata Processing

The CREDO archive is able to extract embedded metadata from some file formats. The archive acquires embedded metadata selectively: from selected file formats, and only selected groups of metadata.

To extract embedded metadata, current version of CREDO uses recognized Apache Tika tool [1]. This tool is called from a REST service that converts the extracted metadata to a simple XML format developed by CREDO (see Fig. 2). The used tool is able to acquire embedded metadata from, among others: PDF documents, image files in TIFF and JPEG formats (EXIF and IPTC metadata), MP3 and WAV audio files (including BWF format [2]), and MP4 video files.

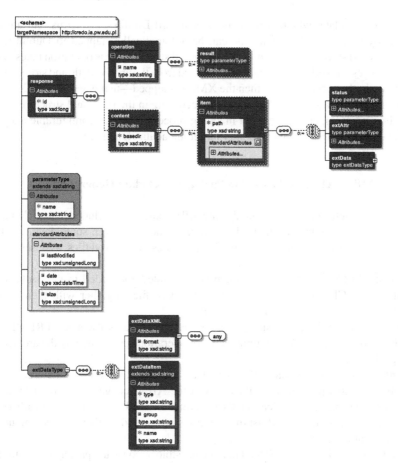

Fig. 2 XML schema diagram of the simple XML format used to transfer and store embedded metadata extracted from delivered files

As any IT tool, the program used for metadata extraction may become unavailable, outdated or may not work in future hardware/software environments. Fortunately, this program is called from the CREDO adapter in BOS (see chapter "CREDO Repository Architecture"), and the adapter provides implementation independent XML-based interface to other subsystems. Therefore the exchange of the extraction program, or even use of more than one program simultaneously, should be quite easy, as it needs only changes in one of services of the adapter.

The extracted metadata can be then stored in the database as XML, or partly converted to metadata stored in generic relational structures.

In certain cases it is necessary to extract some selected values from the metadata, and to store them as discrete elements in the relational structure. This applies primarily to descriptive metadata, like IPTC or embedded Dublin Core metadata.

Some metadata, which correspond to standard Dublin Core terms, but are not explicitly defined as DC elements, can be automatically mapped to Dublin Core elements and stored in this form in the relational structure. This operation takes place in two stages: embedded metadata are extracted from the archived file and stored in the database in the form of XML, then the XML is mapped—using appropriate XQuery expressions—to a set of scalar values for recording in the relational structure.

The extraction, conversion and mapping process is highly configurable and can be adapted to other formats and other target metadata elements.

4.1.4 AIP (Archival Information Package) Metadata Generation

All original metadata files, provided in the SIP package, are included in the AIP package without any changes. The CREDO archive adds, however, some new metadata files. The archive also sets some extended attributes of archived files, as described in Sect. 3.1.

A METS file (in METS 1.6 format) is generated and added to the package. It contains the CREDO package manifest: a list of files together with their digital digests, format specifiers and other metadata.

On the basis of Ingest session journal, CREDO generates also a new PREMIS file (in PREMIS 2.2 format), containing a list of operations performed by the archive on the package.

Information on generated files is added to the package metadata stored in the database, so that periodical verifications of the AIP package will also check these new files. During such verification, the generated PREMIS file is replaced with a new one that contains an updated list of operations performed on the package, including the verification itself.

In future versions of CREDO it may be possible to construct packages of multiple resources, each of which having its own metadata. Additional metadata files may be created in this case.

4.2 Searching Metadata

Metadata of archival objects can be used to search for information in the archive during so-called Search sessions (as stated in OAIS recommendation [4]).

4.2.1 Search Process

Metadata stored in relational structures may be searched by SQL queries, dynamically constructed on basis of user request submitted using end-user application. In future versions of CREDO full-text queries to textual metadata will also be possible.

Metadata stored as XML documents can be searched by explicitly written queries in XQuery language [10].

As a result of the search, packages are found that meet the search conditions. The system lets the user view the metadata of the found packages and select the packages he is interested in. The search request can also be corrected and the search process repeated, until a desired result is obtained.

The search request and result are saved in the database. New Search session may start from a previously saved request. The result of the Search session can also be used as an initial request in later Outgest sessions.

4.2.2   Access Privileges in Search

In the CREDO archive a general rule applies, that an access right to an archived resource, i.e. the possibility of obtaining the resource in Outgest session, have only the resource owner and users representing him.

This rule does not, however, apply to metadata. Access to metadata of archived resources—with the consent of their owners—should be wider.

Access rights to the metadata of an archival package are controlled by so-called access level assigned to the package during the Ingest session. In the CREDO archive four access levels are provided:

- private—only users representing the owner of the archived package have access to the metadata;
- limited—an access to the metadata is granted to users representing indicated customers of the archive;
- internal—an access to the metadata is granted to all registered users of the archive (this is the default level);
- public—the metadata are presented to all archive users and—through an appropriate external infrastructure—to a wide audience.

Ability to search for a particular package, based on its metadata availability, does not mean, however, the possibility of obtaining the contents of the package. The archive customer, not being the owner of the wanted package, should ask the owner for its acquisition and sharing.

4.3   Metadata Processing in Outgest Session

During an Outgest session, a Dissemination Information Package (DIP) is created and made available to the user. The DIP package is stored in the archive buffer, in a dedicated temporary directory created for the session.

The DIP package has a structure of a directory tree, consisted of subdirectories corresponding to archival packages being disseminated. All the metadata files, contained in an AIP package included in the DIP, are copied unchanged to the appropriate subdirectory.

CREDO adds a manifest file (in METS format) to the main DIP directory, containing aggregated information about the structure of the whole DIP package.

4.4 Metadata Processing Configuration

All the metadata processing operations are highly configurable by appropriate records in configuration tables. System administrators can quite easily adjust the list of file formats and associated embedded metadata processing patterns, XML to relational data mappings, file data written to generated METS files, session details written to generated PREMIS files, etc. Addition of a new metadata file format (e.g. XFDU) is more complex, as it requires writing in PL/SQL and registering the code in the configuration tables. But, as the new code will probably base on existing procedures, the addition process is also not very complicated.

Some settings of metadata processing, e.g. list of XML files to be stored in the database, or list of allowable formats together with details of their processing, can be set at customer or even particular session level.

5 Conclusion

The CREDO archive provides means to reliable store and efficiently use the metadata of preserved digital resources. It stores the metadata together with the archived resources in archival packages, so the storage is reliable and it is unlikely to lose the relationship between the metadata and the described resources. It also stores copies of selected metadata in on-line database, so the metadata can be effectively searched and cross-checks can be done between the archival storage and the content of the database.

Flexible data structures and configurable metadata processing methods enable easy adaptation of the CREDO system to new metadata formats and new user's needs.

References

1. Apache Tika. URL http://tika.apache.org. Access: 2016-10-25.
2. British Library Digital Preservation Team: WAV format preservation assessment. URL http://wiki.dpconline.org/images/4/46/WAV_Assessment_v1.0.pdf. Access: 2016-10-25.
3. Consultative Committee for Space Data Systems: XML formatted data unit (XFDU) structure and construction rules. Recommendation for space data system standards. Recommended

standard. (2008). URL http://public.ccsds.org/Pubs/661x0b1.pdf. CCSDS 661.0-B-1. Access: 2016-10-25.

4. Consultative Committee for Space Data Systems: Reference model for an open archival information system (OAIS). Recommended practice. (2012). URL http://public.ccsds.org/pubs/650x0m2.pdf. Access: 2016-10-25.

5. DROID: file format identification tool (2013). URL http://www.nationalarchives.gov.uk/information-management/manage-information/preserving-digital-records/droid. Access: 2016-10-25.

6. Dublin Core Metadata Initiative: URL http://dublincore.org. Access: 2016-10-25.

7. Freed, N., Borenstein, N.: Multipurpose Internet mail extensions (MIME) part one: Format of internet message bodies. RFC 2045 (Draft Standard) (1996). URL http://www.ietf.org/rfc/rfc2045.txt. Updated by RFCs 2184, 2231, 5335, 6532. Access: 2016-10-25.

8. Library of Congress: Metadata encoding & transmission standard. URL http://www.loc.gov/standards/mets. Access: 2016-10-25.

9. PREMIS Editorial Committee: PREMIS data dictionary for preservation metadata (2012). URL http://www.loc.gov/standards/premis/v2/premis-2-2.pdf. Access: 2016-10-25.

10. World Wide Web Consortium (W3C): XQuery 1.0: An XML query language (second edition) (2010). URL http://www.w3.org/TR/xquery. Access: 2016-10-25.

Persistence Management in Long-Term Digital Archive

Piotr Pałka

Abstract The chapter describes the development of persistence management of various data carriers. The goal is to create an abstract system for monitoring and reporting data carriers, providing information about each carrier, a lot of carriers, or area, independently of the carrier type. Further, the abstract system for monitoring solves the problem of lack of appropriate technical solutions for the monitoring and management of many carriers. The approach allows to determine information from which areas to relocate the packages, where they should be relocated, and which areas should be turned on. The information enables the archive to keep high reliability level, and to save the energy.

1 Introduction

The aim of the development of persistence management of data carrier is to create an abstract system for monitoring and reporting data carriers. The abstraction for monitoring and reporting data carriers: (i) is a common mechanism (independent of a carrier type) to provide information about the carrier, or the area (a lot of carriers) to PMS; (ii) is independent of the type of carrier (magnetic tapes, hard disk drives, optical disks (CD, DVD, Blue-ray), holographic disks, minidisks, solid state semi-conductor memory (multimedia cards, USB flash drive,...), etc.); and (iii) solves the problem of the lack of appropriate technical solutions for the monitoring and management for carriers other than hard disks (S.M.A.R.T. technology). Moreover, the abstraction allows to set: (i) from which areas the packages should be relocated, (ii) to which areas the packages should be relocated, and (iii) which areas should be selected to power on (in order to write or read data packets).

P. Pałka (✉)
Institute of Control and Computation Engineering, Warsaw University of Technology, Warsaw, Poland
e-mail: P.Palka@ia.pw.edu.pl

© Springer International Publishing AG 2017   123
T. Traczyk et al. (eds.), *Digital Preservation: Putting It to Work*,
Studies in Computational Intelligence 700, DOI 10.1007/978-3-319-51801-5_7

2 State of the Art

Persistence management is strongly tied with a failure prediction. Papers [4, 8] describe methods for prediction of disk failure. In [8] authors consider a difficult real-world pattern recognition problem, where predicting hard drive failures is done using attributes monitored internally by individual drives. Further, the authors compare the performance of support vector machines, unsupervised clustering, and non-parametric statistical tests. On the other hand, in [8] the abilities of two Bayesian methods to predict disk drive internal failures are investigated. Analyzed Bayesian methods are: (i) naive Bayes submodels that are trained using expectation-maximization, and (ii) naive Bayes classifier, a supervised learning approach. Further, [9] analyzes the data on disk replacement in a large system containing over 100,000 carriers, some for an entire lifetime of 5 years. Authors of the paper find that the annual disk replacement rates typically exceed the nominal annual failure rate, for some systems even tenfold. Schwarz et al. [10] performed similar analysis, they present analysis of disk failure in the Internet archive. They have collected data on disk failure over several years and several installations. The goal is to determine real failure rates. Papers [2, 7] describe the head of the disk analysis, and the impact on operating temperature on the failure prediction. The analysis of impact of the lower head fly-height on the disk reliability is done in [2].

Based on the analysis of literature we can draw the following conclusions. The lifetime of the disk is the longer, the lower the operating temperature of the disk [3, 11]. Secondly, the lifetime of the drive depends on the Power On Hours (length of time, in hours, that electrical power is applied to a device—drives off work longer). Next, a reading failure increases the probability of a media failure [11]. Then, Mean Time Between Failure (MTBF) is directly proportional to the working time, MTBF = [average number of POH (Power On Hours)]/[mean FR (Failure Rate)] [11]. Finally, for hard disks the characteristic called bathtube curve occurs, according to which there are three periods of lifetime: (i) early failure period, for about a year of work, where the hard disk is exposed to high probability of failure; (ii) useful life period, from year to 5–7 years of work, where the device has low probability of failure; and (iii) wearout period, after 5–7 years of work, where a probability of failure grows, and the disk should be replaced.

Schwarz et al. [10] point out the inaccuracy of provided failure parameters, and the user experience results. They explain this as follows. First, the producers do not focus on the historical data, but rather try to determine quality of current products. Second, most quality tests try to artificially accelerate failures, and foresee the hard disks behavior. Moreover, the tests last at best a few months. Third, non-accelerated data quality do not address silent (bit-rot) data corruption. And finally, producers do not take into account disks return.

3   Persistence Management Subsystem

Persistence Management Subsystem (PMS) sets guidelines for the stream layer (file system) on replicas placement, relocation of data, data carriers diagnose and replacement, scheduling the access to the archive, and power management.

3.1   Replication

Replication does not change the bit sequence on the data carrier. It assumes creating multiple copies of the same resource. Additionally, the replication assumes that the resource is copied onto other carriers. In CREDO, there are two levels of replication: low-level replication, that is assured by the file system; and high-level replication, provided with the PMS. Both replication methods serve to ensure data security. Moreover, the high-level replication can be assured on: (i) area level, where different copies of the same resource are kept under different paths, (ii) file system level, where a resource is stored on different file systems, (iii) archive level, where a resource is secured on a few archives.

The higher the level of the replication, the data is safer, as the mechanisms ensure: doubling the resource safety, diversification of a resource representation (when the copies are kept under different file systems), geographical dispersion, when resource is stored in different locations.

3.2   Relocation

The goal of the relocation is to prevent data aging by refreshing the resource on another: carrier, area, file system, or archive. PMS assumes package relocation using optimization methods, where the packages are relocated: periodically, on safer area, with sub-optimal areas usage, and with energy efficiency. Moreover, the relocation is coupled with the carriers replacement. The packages are automatically high-level replicated among the regions, the PMS schedules the plan for a long-period.

3.3   Data Carriers Diagnose and Replacement

The goals of the diagnose and replacement of data carriers module are:

- Analysis of a risk concerning failures of data carriers and whole areas designed for storage of archival packages. The analysis consider single pieces, lots, models of carriers, and it goal is to predict the failure, and secure the data stored on the carrier by replication, relocation on the other carriers, renewing, or emptying it.

- Replacement of carriers decision support system.
- Failure prevention due to prediction of the failure.
- Interoperability with the data relocation, by automatic rewriting data from the carriers under threat.

To obtain the goals, the module for carriers, and area reliability assessment is used. It is common mechanism to provide the information about different: carriers, areas, batches of carries to PMS. Moreover it provides information independently of carrier type (hard disk, magnetic tape, CD, DVD, Blue-Ray, pendrive, etc.). Finally it solves the problem of a lack of technical solutions for monitoring, and managing carriers. The module allows to choose a source or destination for the archival packages during archive operation. Also, it specifies the moment for the relocation. Finally, it determines the areas for powering-on during reading or writing the archival packages.

3.4 Power Management

In CREDO the carriers are grouped into storage areas, that are subject to the PMS management. Single storage area has assigned a number of data carriers. The allocation of the packages onto storage areas is done by PMS, and a power management module manages areas starting-up, and shutting-down. Power management system is coupled with a scheduling module, and it has to start-up given area, when the schedule assumes operating on it. Similarly, when planned operations finish, the power management module shuts the area down.

4 Data Carriers Diagnose and Replacement Module

The goal of the diagnose and replacement of data carriers module is to create the generic mechanism handling different media, diagnosing them, and providing the failure predicting measures. The measures do not depend on specific media type.

4.1 Measures of Reliability

With respect to each carrier, a layer of abstraction for the monitoring and reporting of carriers, returns following measures.

- The reliability percentile, that determines the probability that the carrier will be reliable within a time. In other words, the Xth reliability percentile has the value of Y, when the probability that the carrier is reliable over the next Y days is equal to $X/100\%$. For the reliability percentile following inequalities hold:

$$P((-\infty, Y]) \geq X/100 \tag{1}$$

and

$$P([Y, \infty)) \geq 1 - X/100 \tag{2}$$

- The probability of data carrier failure within 1 year.

4.2 Failure Analysis

MTBF (Mean Time Between Failures)

MTBF is an indicator used to determine the viability of hard drives. This is the average time expressed in hours, during which the device can operate without interruption (failure). Manufacturers carriers provide this parameter [5]. Parameter MTBF is the sum of the MTTF and MTTR: $MTBF = MTTF + MTTR$.

MTTF (Mean Time To Failure)

MTTF indicates average operating time from the start of operation or from its last repairs to the establishment of the first crash. For the data carriers we assume that $MTTF = MTBF$.

MTTR (Mean Time To Repair)

MTTR determines the average time from the moment of failure to repair damaged equipment [6]. For the data carriers we assume that $MTTR = 0$, as the data carriers most often are not reparable.

4.3 Reliability

Reliability is a property of the object, that tells about whether it works correctly (meets all the assigned functions and activities) for the required time and under certain conditions (in the assembly forcing factors).

Reliability of the object is determined by the probability of the event described by:

$$R(t) = Pr(T > t) = \int_t^{\infty} f(x)dx \tag{3}$$

where $R(t)$ is a reliability function, t is time of proper working of the object (without any failure), and T is assumed (or required) time of failure-free working.

For unrecoverable objects (as most data storage devices are), we assume following:

- for $t = 0$, $R(0) = 1$—in current moment the object works properly,
- reliability is non-increasing function,
- for t tending to infinity $\lim_{t \to \infty} R(t) = 0$, which means that once every object is going to break down.

Linking the reliability function with the cumulative distribution probability of failure leads to:

$$R(t) = Pr(T > t) = 1 - Pr(T \leq t) = 1 - F(t) \tag{4}$$

With the reliability function, the parameter MTBF is related with the following relationship:

$$R(t) = e^{-t/MTBF} \tag{5}$$

For example, for Seagate Barracuda Green 2 TB ST2000DL003 hard disk, MTBF parameter is equal to 750,000 h. The probability for working without any failure for 1 year of the disk is equal:

$$CAFR = 1 - R(365 * 24) = 1 - e^{-365 * 24/750,000} = 1 - e^{-0.01168} = \tag{6}$$
$$= 1 - 0.011612 = 0.988388$$

namely, the probability of failure within one year is equal 0.011612, or 1.1612%. For the data carrier, the 99th percentile of an area failure probability is equal to $CFP^{99} = 314$ days.

5 Implementation

To deal with the failure estimation, we assume use of the reliability concept, that describes the ability of a system or component to function under stated conditions for a specified period of time. The reliability defines the probability of proper working defined by a function: $R(t) = P(T > t) = \int_t^\infty f(x)dx$, where t is the reliable working time, T is the assumed time for reliable working, and $R(t)$ is the reliability function. The reliability function is coupled with the MTBF parameter: $R(t) = e^{-t/MTBF}$. We propose the use of the reliability function $R(t)$ to calculate the distribution function of work of data carrier and on this basis percentiles. We suggest the following equations to calculate the percentiles $P(x)$ (x is degree of percentile).

We developed the method to predict the failure in situation where we have only basic information on a media (e.g. only NARA guidelines, and MTBF), or we have specific monitoring and diagnosing tool, as S.M.A.R.T.

1. With only the NARA guidelines available, we take the admissible operating media time as MTBF parameter.
2. With the MTBF available, we calculate the reliability function: $R(t) = e^{-t/MTBF}$, where t is the time (in hours), that passed from the carrier manufacturing. The percentile is calculated using following equation: $P(x) := -ln(x) * MTBF$.
3. Apply the carrier-specific reliability recalculation—for each of carrier type, there should be developed carrier-specific implementation recalculation, and the goal is to obtain most proper reliability setting.

Having the reliability function, we can formulate the measures for reliability for each of the data carriers. We propose two measures:

- 99th percentile of carrier failure probability: $CFP^{99} = -ln(0.99) * \frac{MTBF-t}{24}$,
- probability of annual failure rate: $CAFR = 1 - e^{\frac{-365*24}{MTBF-t}}$.

5.1 Hard Disk Specific Reliability Calculation

The specific implementation is proposed for the hard disks where the information from the S.M.A.R.T. monitoring system can be used.

1. With the S.M.A.R.T. parameter no. 009 (Power On Hours)—POH, the reliability function is more accurate: $R(t) = e^{-POH(t)/MTBF}$, where $POH(t)$ is the value of POH in the current moment.
2. With the S.M.A.R.T. parameter no. 194 (Temperature Celsius)—TC, we can modify the working time of the disk, according to Arrhenius equation, that takes into account dependence of failure and working temperature. As the MTBF parameter is set in the reference temperature TC_{ref} (mostly 30 °C), having the information on working temperature, we can calculate the AF factor, that is used to modify the working time.
To calculate the modifier of working time, we apply the Arrhenius equation, that ties the reliability with the working temperature:

$$Arh(TC) = A * e^{\frac{E}{k*TC}} \tag{7}$$

As the MTBF parameter is set in reference temperature TC_{ref} (usually 30 °C), and the disk may work in different temperature, we can calculate the rate of failure change:

$$AF = \frac{R_{TC_{ref}(t)}}{R_{TC}(t)} = e^{\frac{E}{k}*\left(\frac{1}{TC_{ref}} - \frac{1}{TC}\right)} \tag{8}$$

where:

- E—disk head activation energy (for default $E = 0.6$ eV),
- k—Boltzmann constant, $k = 8.6173303(50) \times 10^{-5}$ ev/K,
- TC—Celsius working temperature for hard disk in Celsius degrees,

- TC_{ref}—reference temperature in which the MTBF parameter was set,
- A—Hamaker constant, $A \approx 10^{-19}$ J (on the base of [1]).

Having calculated AF parameter, we change "the time elapsed" since the last reading of the modified parameter TC. If we read the temperature in 1 h, and the parameter $AF = 2$, then we change the elapsed time in 2 h instead of 1. This requires calculating the passage of time for each data carrier.

3. If we have the S.M.A.R.T. parameters about the time the drive (S.M.A.R.T. parameter 009—Power On Hours), then we can calculate how much in fact the hard disk is working (this is important if the drives are usually in off mode) [11]. Having given this parameter, then instead of how much time actually passed from the manufacture of disk space, calculate the time based on the parameter POH: $\Delta t := \Delta POH * AF$, where Δt is the time disk is calculated from the last access to S.M.A.R.T. parameters.

4. If we have the S.M.A.R.T. parameters on the number of start/stop of the drive (S.M.A.R.T. parameter 004—Start Stop Count) then we can estimate the impact of start/stop on the disk operation time. Having given this parameter and parameter POH, then instead of how much time actually passed from the manufacture of disk space, we calculate the time based on parameters POH and SSC: $\Delta t := (\Delta POH + \Delta SSC * T_{SSC}) * AF$ where T_{SSC} multiplier specifying age your disk as a result of its on/off cycles.

5. Literature review gives us some clues, which S.M.A.R.T. parameters warn us of impending failure. When one of the parameters: Read Error Rate (S.M.A.R.T. 001), Reallocated Sectors Count (005), Seek Error Rate (007), Spin Retry Count (010), Calibration Retry Count (011), Reported Uncorrectable Errors (187), Command Timeout (188), Current Pending Sector Count (197), or Offline Uncorrectable (198) occurs, the probability of failure grows rapidly. Therefore, we modify the MTBF parameter as follows: $MTBF \leftarrow MTBF - \frac{\Delta P}{ADM\_P}$, where ΔP is the change of the S.M.A.R.T. parameter, and ADM_P is admissible value of this parameter.

5.2 Persistence Parameters for Areas

To aggregate the measures for areas of carriers, we propose the min and max operators (A is a set of carriers c belonging to an area):

- 99th percentile of an area failure probability: $AFP^{99} = \min_{c \in A} CFP_c^{99}$,
- probability of annual failure rate: $AAFR = \max_{c \in A} CARF_a$.

The module allows to choose a source or destination for the archival packages during archive operation. Also, it specifies the moment for the relocation. Finally it determines the areas for powering-on during reading or writing the archival packages.

Having reliability measures calculated, we can choose a source or destination for the archival packages during archive operation, by selecting the destination areas with lower *AAFR* value, and source area with higher *AAFR* values. Also, the moment for relocation is specified on the base of AFP^{99} measure. Finally, the obsolete areas are determined on the base of AFP^{99} measure.

6 Durability Management Methods in Long-Term Archive

Durability management methods in long-term archive can be divided into: (i) planning operation for packages maintenance, carriers and areas in the archive (magnetic refresh periodic recalculation of checksums, high level relocation); (ii) assessment of areas in terms of reliability; and (iii) setting obsolete carriers/areas.

Scheduling

To effectively plan maintenance operations, it is proposed to use a 99th percentile for area reliability AFP^{99}, as percentile indicates a time for which the probability of failure is equal to 0.01. To protect the data on the area before the crash, should be at $\frac{1}{K}AFP^{99}$ days to carry out operations: magnetic refresh periodically calculate checksums, as well as perform a high-level relocation. To ensure correct operation of the module reliability, value K should be determined by an expert.

Area Reliability Coefficients

Reliability parameters provide us with knowledge of the "quality" of carriers and areas. We can determine, among others, area reliability coefficients that are used to select areas in the allocation of packages.

Obsolete Media Determination

Outdated carriers are those, that carry very high probability of failure, and should be replaced with newer ones. For these carriers there are no plans to write new packages, and shall not be designated as a target of high-level relocation. It is proposed to recognize that the carrier is outdated, using the 95th percentile of carrier reliability. Value of the percentile less than D days: $CFP^{95} < D$, means that the carrier is outdated. To ensure correct operation of the reliability module, value of D should be determined by an expert.

7 Conclusion

Concluding, described subsystem for persistence management of different data carriers was implemented in CREDO Persistence Management Subsystem. The abstraction for data carriers diagnose and replacement module was also implemented, and one specific module for hard disk reliability calculation using S.M.A.R.T. technology was also developed and implemented, proposing specific calculations that applies different S.M.A.R.T. measures. Next, we propose the generic measures to define the persistence for both particular carriers, and whole areas. Finally, the utilization of the persistence indices in long-term archive is described: to support scheduling, replacing the media just before the breakdown, packages allocation.

References

1. Ambekar, R., Gupta, V., Bogy, D.B. Experimental and numerical investigation of dynamic instability in the head disk interface at proximity. *Journal of Tribology*, 127(3):530–536, 2005.
2. Elerath, J.G., Shah, S. Disk drive reliability case study: dependence upon head fly-height and quantity of heads. *Reliability and Maintainability, 2003 Annual Symposium-RAMS*, pp. 151–156. IEEE, 2004.
3. Elerath, J.G., Shah, S. Server class disk drives: how reliable are they? *Reliability and Maintainability, 2004 Annual Symposium-RAMS*, pp. 608–612. IEEE, 2004.
4. Hamerly, G., Elkan, C., et al. Bayesian approaches to failure prediction for disk drives. *ICML*, pp. 202–209. Citeseer, 2001.
5. http://en.wikipedia.org/wiki/Mean_time_between_failures, access 2016-10-17.
6. http://en.wikipedia.org/wiki/Mean_time_to_recovery, access 2016-10-17.
7. Mao, S., Chen, Y., Liu, F., Chen, X., Xu, B., Lu, P., Patwari, M., Xi, H., Chang, C., Miller, B., et al. Commercial TMR heads for hard disk drives: characterization and extendibility at 300 gbit/in 2. *Magnetics, IEEE Transactions on*, 42(2):97–102, 2006.
8. Murray, J.F., Hughes, G.F., Kreutz-Delgado, K. Hard drive failure prediction using non-parametric statistical methods. *Proceedings of ICANN/ICONIP*. Citeseer, 2003.
9. Schroeder, B., Gibson, G.A. Disk failures in the real world: What does an MTTF of 1,000,000 hours mean to you? *FAST*, vol. 7, pp. 1–16, 2007.
10. Schwarz, T., Baker, M., Bassi, S., Baumgart, B., Flagg, W., van Ingen, C., Joste, K., Manasse, M., Shah, M. Disk failure investigations at the internet archive. *Work-in-Progress session, NASA/IEEE Conference on Mass Storage Systems and Technologies (MSST2006)*, 2006.
11. Yang, J., Sun, F. A comprehensive review of hard-disk drive reliability. *Reliability and Maintainability Symposium, 1999. Proceedings. Annual*, pp. 403–409. IEEE, 1999.

Power Efficiency and Scheduling Access to the Archive

Tomasz Śliwiński

Abstract Available storage systems offer different levels of interplay between data security and long term storage costs. The archive management system must ensure not only the best possible values of both criteria, but also allow to specify decision maker preferences regarding their importance. The chapter describes in detail the problem of managing hardware and data in the CREDO archive system. It elaborates on needed assumptions and presents sophisticated scheduling algorithm for the management of archive operations.

1 Introduction

As the repository lifetime is very long, the data carriers cannot be on all the time. On contrary, the carriers will be off most of the time, and will be powered only when needed. In the repository using magnetic data tapes, the carriers should be loaded/unloaded into/from the reader if needed. Access to the data stored in the archive requires powering on the appropriate carrier, or loading the appropriate tape into the reader. Effective and reliable usage of the archive requires separate algorithms to schedule access to its resources, and to manage turning on/off or loading/unloading data carriers. The needed data structures and algorithms constitute separate software module—the scheduling module. Developing data and hardware management algorithms requires proper identification of sources of storage costs and safety hazards.

The costs can be divided in two categories:

- power supply and maintenance costs,
- the cost of the hardware itself (wear).

In the case of the hard drives, simple arithmetics shows the cost of hardware is comparable to the cost of the power supply over the whole live span of the drive, assuming the drive operates permanently. It is hard to be estimated if the drive can

T. Śliwiński (✉)
Institute of Control and Computation Engineering, Warsaw University of Technology,
Warsaw, Poland
e-mail: T.Sliwinski@ia.pw.edu.pl

© Springer International Publishing AG 2017   133
T. Traczyk et al. (eds.), *Digital Preservation: Putting It to Work*,
Studies in Computational Intelligence 700, DOI 10.1007/978-3-319-51801-5_8

be switched off for some periods. Considering the magnetic tapes, the hardware cost is definitely higher than the cost of the power supply. The maintenance cost is hard to estimate, and strongly depends on the management system—the way it supports administrators in their daily operations.

The power supply cost results from the turning on/off and from normal operation of the device. For hard drives, the start up costs are negligible as compared to cost of continuous operation, but this is not necessary true for other media types.

The cost of equipment is determined by its purchase price, but can be spread over a long period of time, depending on the usage patterns. Typical hard drive, operating in 24/7 regime, has a mean time between failures (MTBF) of 500,000 to 1,000,000 h, and the real data shows annual failure rate between 1 and 10 % (see [4]). Different usage patterns are possible. The drive can operate all the time, but also can be spinned off (using internal drive controller), or powered off completely by cutting off its power supply. In all cases, one has to keep in mind the manufacturer limitation of the number of load/unload cycles, that is around 300,000 for most drives. This however, does not seem to be a serious limitation, as this, assuming 10 year operation period, translates to approximately 80 load/unload cycles a day. Estimating cost of hardware is particularly difficult for hard drives that do not operate all the time, but are switched off for longer periods. There are two sources of estimation errors—one results from the fact that all reliability tests are performed for drives that operate all the time, and there is no method to extrapolate them to the mixed usage patterns, and the other results from the fact that hard drives require periodic renewal of the magnetic domains that are used to store the data. This procedure is performed by the drive's software during its idle state, but cannot be performed when the drive is not spinning. Again, there are no statistics, concerning reliability of data in such cases, and no recommendations concerning lengths of those power off periods.

The tapes are meant for off line data storage, but lack versatility and data accessibility. Although, they are claimed to last for up to 30 years, the more realistic lifespan for magnetic tape is around 10 years, and to ensure maximum data safety, not more that 4 years. For the tapes the most limiting factor is the number of rewinds, approx. 150, before mechanically degrading the physical medium. Apart from obvious disadvantages in the form of long access times, the tape systems also limit the number of tapes loaded in parallel, based on the number of physical tape drives. The costs in tape systems come from buying the tape system itself (automatic tape library) with tape cartridges, and from the need to exchange failed cartridges and/or drives with new ones. But the power supply are not the major component of total usage costs.

In the CREDO archive system data reliability is ensured by the PMS (Persistence Management Subsystem) module. Its main objectives are: monitoring of archive storage (data verification, hardware diagnosis), making recommendations for data allocation and relocation, power management and scheduling the archive operations. The data reliability is secured with the following mechanisms:

- low-level copies,
- high-level replicas,
- periodic data checks,

- monitoring and management of data carriers with estimation of their reliability,
- estimation of the reliability of the areas.

Managing low-level copies lies outside of the PMS, but PMS and scheduling algorithms must be aware of the real number of copies as they induce the real data reliability level. High-level replicas are meant to occupy different physical locations to make them less prone to big scale events, like disasters, floods, fires, etc. They also can use different underlying technologies, better suiting the customer needs. Periodic data checks are crucial for preserving data integrity, and the interval between consecutive checks, has big influence on final estimation of data reliability. As stated in chapter "Persistence Management in Long-Term Digital Archive", the monitoring and management of data carriers is meant to provide additional abstract layer (common interface) for collecting information on the current state of each single data carrier. The aggregated parameters can be then used to estimate the reliability of the whole areas.

The administrator and/or higher level decision maker can differentiate their products based on various cost/reliability ratios. This can be done mainly by setting different numbers of high-level replicas and low-level copies, that consume different amount of available storage space, but significantly improve the data reliability. For the same reason, the decision maker can alter the frequency of data integrity checks, influencing the power supply costs and wear of the hardware.

2 Problem Description

In this section the detailed description of the problem is given, from the viewpoint of the optimization algorithm.

There is given a set of operations that are to be scheduled. Operations belong to procedures. Single procedure consists of a sequence of sets of operations. The sequence defines the order in which the sets are executed. All operations within a set can be run in parallel (see Fig. 1).

Each operation has defined a time period within which it has to start and finish. The starting time of the operation can be given in advance or be undefined, in the latter

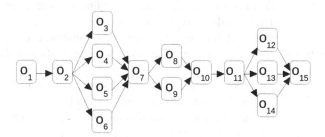

Fig. 1 Example of the procedure structure

case it will be a result of the scheduling algorithm. Some of the operations require access to storage areas, which constitute critical resources. In particular, the size, the number of parallel reads, and writes are limited. Additionally, some operations require exclusive access to the resource (area) either for write only, or for read and write.

Area is the smallest indivisible part of the archive of known capacity that can be turned on and off. Area may consist of a number of data carriers, but they are indistinguishable by the scheduling algorithm. However, their parameters can affect some aggregated parameters of the area.

For each operation there is given a set of source areas (to read data from) and a set of target areas (to write to or to modify data). In the final schedule, only one area out of the given set can be chosen to read data from and/or one area to write data to.

The scheduling module should create the optimized schedule meeting the following multiple objectives (multiple criteria): ensuring data security and integrity, minimizing usage cost of the archive, load and I/O traffic leveling between different parts of the archive (areas).

2.1 Notation

Let i denote operation ($i \in I$), and p denote procedure ($p \in P$). Set of operations belonging to procedure p is denoted by I_p. Some operations utilize areas o ($o \in O$), and each such action has defined separate set of possible areas for reading $U_i \subseteq O$ and writing data $S_i \subseteq O$. As noted earlier, for each operation i time period (T_i^e, T_i^l) is given within which the operation has to start and finish. The starting time of the operation is denoted by T_i^s. The maximum number of parallel operations (read and write) in area o is denoted by R_o, and parallel write operations by W_o.

Operations within a procedure are grouped to create sets of operations that can be run in parallel. There is a defined order of execution of such sets, and G_p^k denotes the set of operations that can be executed simultaneously in position k, where $k = (1, \ldots, K)$.

Below is the list of all operation parameters relevant the scheduling algorithm.

Predecessors	Set of operations required to start/finish before given operation. The precedence type is defined later.
Lead lag	Required time period before given operation can start.
Precedence type	There are two precedence types relevant to the scheduling algorithm:
FS	Operation can start only after all the operations in the Predecessors set have finished, plus given Lead lag.
SS	Operation can start only after all the operations in the Predecessors set have started, plus given Lead lag.

T^e, T^l	Time period in which the operation has to be executed, i.e. cannot start earlier than T^e and finish later than T^l.
Source areas	Set of areas to read data from. Only one will be chosen in the final schedule.
Read times	For each source area the estimated reading time.
Target areas	Set of target areas. Only one will be chosen to write to.
Write times	For each target area estimated write/modification time.
Write size	For each target area estimated size of the data written—due to the file system properties can be different for each area.
Lock type	Type of lock put on the target area to during execution of the operation. Possible lock types are:
Full	No other operation on the area allowed.
Write	Only read operations are allowed.
None	No lock. All operations are allowed.
D^{min}, D^{max}	Minimum and maximum duration, defined for those operations which do not use resources (areas). The scheduling algorithm assigns the actual duration from this interval according to some strategy.
T^s	Starting time of the operation. This is input and output data for the algorithm. If it is not set, or beyond (T^e, T^l), then the scheduling algorithm sets its value.

The list of area properties considered by the scheduling algorithm is as follows.

Size	The maximum number of bytes that can be stored in the area. In different file system, the actual size available for data can be different (smaller).
Used space	Number of bytes used by file system to store all the data.
R_o	Maximum number of I/O operations performed in parallel in the area o.
W_o	Maximum number of write I/O operations performed in parallel in the area o.
Power off time	Minimum time period with no operations, that results in area powering off in order to save energy.
Power up cost	Total cost of powering on the area.
Power on cost	Cost of the area operation in the time unit.
Power on time unit	Time unit for which the area operation cost is defined.
Reliability factor	Some numerical quantification of area reliability.

The separate problem is acquiring information about the actual space used by the stored data on each area. This cannot be done during execution of write/modify operations. Therefore, a special write blocking operation will be provided only to read the used space of the area.

2.2 Basic Assumptions

During development of the scheduling module the following assumptions were made.

1. Multiple optimization criteria have to be considered.
2. Areas can be grouped in regions, each region is in possibly distinct geographical location.
3. To ensure data integrity, multiple (at least two) high-level replicas are created, each copy on separate region.
4. All the optimization criteria should be considered when choosing areas to perform given operation on. In particular, each operation can read data from one of the high-level replicas and write data to one of the areas belonging to the given region.
5. It is assumed, the execution time of each operation in a set of simultaneously executed operations can be estimated in advance. If it cannot be done, the scheduling of the following sets of operations in the procedure is suspended, i.e. will not be scheduled.
6. It is assumed that the execution time of each operation does not depend on the total number of operations executed in the same time period. Due to the physical properties of the file system, this assumption obviously does not apply to the real systems in every usage pattern. In most systems, however, this assumption can be made if the number of parallel operations is limited.
7. For operations that use both, source and target resources, they are used in exactly the same time interval.
8. Previously created schedule in the presence of new operations (incremental scheduling), should stay unchanged as long as no constraints are violated.

3 Scheduling Algorithm

The algorithm is based on the construction heuristics (see [1, 7]). It builds the final solution using greedy method for operations sequenced according to some criteria.

Let τ be a permutation of a set I defining initial sequence of operations, i.e. if $i < j$, then operation $\tau(i)$ is placed before operation $\tau(j)$ in the sequence. Let $\tau(1) \ldots \tau(n-1)$ be operations used to create some partial, already existing schedule and the algorithm is about to schedule operation $\tau(n)$.

Let *event* be any change of the system state in a partial schedule, i.e. start or finish of any operation that requires resources (areas). All events in a partial schedule can be ordered according to their times: $t_1 < \cdots < t_m$.

The algorithm of scheduling operation $\tau(n)$ works as follows.

1. The operation that does not use resources is scheduled as soon as possible, but only as long as it does not violate the precedence relations in a procedure.
2. Operation that uses resources is tentatively scheduled in each time $t_k \in \{t_1 \ldots t_m\}$, in a way that either operation starts or finishes in time t_k.

3. For each time $t_k \in \{t_1, \ldots, t_m\}$ all source areas and target areas are considered for usage.
4. The algorithm selects time $\bar{t}$ (either as operation's start or finish), the single area from the set of source areas and/or single area from the set of target areas.
5. In the above selection the algorithm tries to minimize the increase of the objective value *marginal cost* resulting from execution of the operation. The objective here is an aggregation of the optimization criteria (see detailed description later).
6. Only those solutions are considered (times and resources/areas) for which no constraints are violated (precedence within procedure, area capacity, maximum number of parallel operations).

3.1 Initial Sequence of Operations

As a greedy construction heuristic is used and no backtracking is possible, proper initial sequence is crucial for obtaining good final schedules. The general principle here is first to sequence operations with lower degree of freedom (lower T^l, lower cardinality of source/target area sets, etc.) and later operations with higher degree of freedom. This way we maximize the likelihood for achieving good final schedules. The sequence should also meet the precedence requirements of operations within the procedures, that is operations that are sequenced later in a procedure should also be positioned later in the initial sequence.

The exact strategy of ordering the operations works as follows. Let d_i be estimated duration of operation i and a_i the total cardinality of source areas set and target areas set. The duration of the operation can be deliberately undefined for some operations. Let operator *def* return logical true if and only if its argument is defined and operator *undef* return logical true if and only if its argument is undefined. Operation i will be sequenced before operation j only if

- $def(T_i^s)$ and $undef(T_j^s)$ or
- $def(d_i)$ and $undef(d_j)$ or
- $def(d_i)$ and $def(d_j)$ and $def(T_i^s)$ and $def(T_j^s)$ and $T_i^s + d_i < T_j^s + d_j$ or
- $(undef(d_i)$ or $undef(d_j)$ or $undef(T_i^s)$ or $undef(T_j^s))$ and

  - $def(T_i^l)$ and $undef(T_j^l)$ or
  - $def(T_i^l)$ and $def(T_j^l)$ and $T_i^l < T_j^l$ or
  - $def(T_i^e)$ and $undef(T_j^e)$ or
  - $def(T_i^e)$ and $def(T_j^e)$ and $T_i^e < T_j^e$ or
  - $a_i < a_j$ or
  - $def(d_i)$ and $def(d_j)$ and $d_i > d_j$

The above sequencing strategy will by further denoted by the operator $ord(i, j)$, which returns true only if operation i should be sequenced before j.

To take into account the precedence of operations within the procedure, the operator $ord(i, j)$ is embedded into the following algorithm.

Step 1 $m := 0$, $S_0 := \emptyset$ (where S is the set of sequenced operations).

Step 2 Let C be the sum of sets G_p^k, for all procedures $p \in P$ and k indicating the first set in a sequence of sets within procedure p in which not all operations have yet been sequenced, i.e. $C = \bigcup_{p \in P, \min k: G_p^k \cup S_m \neq \emptyset} G_p^k$. The best operation $i \in C$ is selected, i.e. $i \in C : \nexists_{j \in C} ord(j, i) = $ true. Add i to set S: $S_{m+1} := S_m \cup \{i\}$. $m := m + 1$. Unless $C = \emptyset$, repeat Step 2.

4 Optimization Objective

The following objectives are considered by the scheduling algorithm.

1. Total operation cost of the archive within the scheduled time period.
2. Reliability factor.
3. Relative area space usage.
4. Relative I/O operations usage.

All criteria, except the second one, are to be minimized. In the following, we give more details on the criteria, and present the scalarization required to use them within the optimization procedure.

4.1 Total Operation Cost

Each area has the following cost related parameters.

- Power up cost—total cost of powering up the area.
- Power on cost—total cost of running the area in a defined unit of time.

For each operation depending on its placement in the schedule and used areas (source and/or target) the algorithm computes the total operation cost being the sum of the cost of powering it up (if the area is being actually powered up) and the running cost (if the running time increases when compared to the existing schedule). For each area additional 'Power off time' parameter is specified. It defines minimum time period without any operations which causes the area to be powered down to save energy. As data carriers can have different physical properties the reliability of an area can be partially affected by proper setting the above parameters. For example, one can increase the Power up cost and Power off time to minimize the number of area's on/off operations.

4.2 Reliability Factor

For each area some reliability factor is defined and areas with higher reliability are preferred, to increase data security and integrity.

4.3 Relative Area Space Usage

Based on the area size, used space and the size change made by the operation the relative space usage of an area is determined. The algorithm tries to keep this value at similar level on all areas, preferring areas with lower values of this objective.

4.4 Relative I/O Operations Usage

Similarly, the number of actual I/O operations in relation to the maximum number of I/O operations is leveled between areas by the algorithm. This should prevent the situation where some areas are extensively utilized leading to bottlenecks.

4.5 Scalarization with Preference Expressing Weights

To make the multicriteria problem computationally tractable, one needs to aggregate the individual criteria to generate efficient solutions. There are many different aggregation methods, resulting from different approaches multicriteria optimization. The examples are: weighted sum, maximin of the objectives, reference point method, etc. (see [2, 3, 5]). Due to the properties of the construction heuristic we utilized the basic weighted sum aggregation.

The four criteria utilized here refer to different parameters of the schedule with different physical or logical interpretation. Furthermore, in the construction algorithm utilized here there is no need to know global objective, or even the absolute value of the marginal cost. It is sufficient to be able to evaluate relative differences among criteria for all examined potential solutions. Let $y = (y_1, y_2, y_3, y_4)$ and $z = (z_1, z_2, z_3, z_4)$ be values of the four criteria (objectives) under consideration for two solutions, respectively.

Let the vector $w = (w_1, w_2, w_3, w_4)$, $\sum_{i=1}^4 w_i = 1$ of normalized weights express decision-maker preferences for individual criteria.

In the approach applied in the scheduling module the solution represented by vector y is better than solution represented by vector z if and only if

$$w_1 \frac{y_1 - z_1}{\max\{y_1, z_1\}} + w_2 \frac{z_2 - y_2}{\max\{y_2, z_2\}} + w_3 \frac{y_3 - z_3}{\max\{y_3, z_3\}} + w_4 \frac{y_4 - z_4}{\max\{y_4, z_4\}} < 0$$

Note the opposite sign of the fraction for criterion 2 which is maximized. If both solutions are equal (no one is better), then the solution with operation starting earlier is preferred.

One should note, that the above approach does not guarantee finding global Pareto optimal solution [6]. It performs only local search for optimal solution by repeating the scheduling procedure for each operation taken from the initial sequence.

5   Conclusion

The CREDO archive system requires highly customizable and efficient scheduling algorithm. The developed optimization procedure takes into account multiple real life constraints and requirements, and still allows expressing different preferences regarding data safety, reliability and power efficiency. The algorithm goes well beyond simple construction heuristics by introducing specialized initiation procedures, objectives, and taking into account multiple areas in a single pass. It has demonstrated its usefulness as a heart of the complex archive management system.

References

1. Braesel, H., Herms, A., Moerig, M., Tautenhahn, T., Tusch, J., Werner, F., Heuristic constructive algorithms for open shop scheduling to minimize mean flow time, *European Journal of Operational Research*, 189(3):856–870, 2008.
2. Ehrgott, M., Multicriteria Optimization, 2005, Springer.
3. Gandibleux, X., Multiple Criteria Optimization: State of the Art Annotated Bibliographic Surveys, Springer, 2002.
4. Klein, A., Q1 2016 Hard Drive Stats, Backblaze 2016, http://www.backblaze.com/blog/hard-drive-reliability-stats-q1-2016.
5. Legriel, J., Multi-Criteria Optimization and its Application to Multi-Processor Embedded Systems, Ph.D. Thesis, Universite de Grenoble, 2011.
6. Ogryczak, W., Equity, Fairness and Multicriteria Optimization, *Multiple Criteria Decision Making*, University of Economics in Katowice, 1:185-199, 2006.
7. Semančo, P., Modrák, V., A Comparison of Constructive Heuristics with the Objective of Minimizing Makespan in the Flow-Shop Scheduling Problem, *Acta Polytechnica Hungarica*, 9(5):177-190, 2012.

Information Management in Federated Digital Archives

Piotr Pałka and Tomasz Traczyk

Abstract The chapter describes the problem of resources dislocation among federated archives. In the archive lifetime, the resource can be a subject of format change, relocation, transfer onto another operating system or area, the data carrier can be a subject of magnetic renewal, etc. Some of the operations can be risky, namely the resource can be, in the effect of them, damaged. The federation of archives allows for dislocation of digital resources, i.e. the another method to ensure reliable ability to read it. As the federated archives are usually geographically dispersed, have different operating systems, or storage media, are managed in various ways, the dislocation prevents against the effects of natural disasters, acts of war, etc. The chapter describes acquisition and placement of information about the corresponding resources, archives identification, and communication among the archives. The developed coordination protocol, basing on the state of the art, is described and presented under the assumption, that the archives communicate as they have software agents to communicate with each other. It allows among others, that the archives do not need to be homogeneous. Finally, the implementation proposal of the module responsible for relocation is described.

1 Introduction

Dislocation of digital resources is one of the methods of ensuring reliable ability to read digital data. Because it is the only known way to protect resources from the effects of natural disasters, acts of war, etc., it must be recognized that in professional digital repository possibility of resources dislocation is essential. Main goal of replication is the possibility of semi-automatic reconstruction of the damaged resource.

P. Pałka (✉) · T. Traczyk
Institute of Control and Computation Engineering, Warsaw University of Technology,
Warsaw, Poland
e-mail: P.Palka@ia.pw.edu.pl

T. Traczyk
e-mail: T.Traczyk@ia.pw.edu.pl

© Springer International Publishing AG 2017
T. Traczyk et al. (eds.), *Digital Preservation: Putting It to Work*,
Studies in Computational Intelligence 700, DOI 10.1007/978-3-319-51801-5_9

The possibility is semi-automatic, because an operator decides on replacement of a damaged copy of a dislocated copy, and means of metadata manipulation.

Mentioned federation of archives is a group of geographically dispersed archives that may communicate and cooperate with each other. Those archives may be specialized in terms of applications (e.g. official, film, music, television, medical, etc.), may have different operating system, filesystem, and/or storage media. Moreover, federated archives can have different management systems, or organizational structure. Nevertheless, the main issue is possibility of communication, ensuring among other eventuality of digital data transfer.

The federation of archives provides, among others, the issue of digital resources dislocation. Digital resources dislocation is one of the method of ensuring a reliable ability to read digital data. The goals are as follows.

- The digital data reliability due to existence of distant copies. When in one archive package is destroyed, the archive can enquire to the other archive for a transfer of corresponding package.
- Organizational and ownership diversification. Two archives may have different ownership and organizational structure, resulting with different ways of package reliability ensuring.
- Format diversification of the same data. The same data can exist in different formats, according to e.g. different format transformation policies, providing protection among the format expiration.
- Application diversification (e.g. deep or shallow archive).

There exist at least two kinds of dislocation of digital resources.

- Replication within a single archive in remote filesystems. It ensures bitstream preservation of resources.
- Replication in federated archives. The bitstream identity is not required.

1.1 Replication Within a Single Archive

Each digital resource can be stored in several high-level replicas, which can be kept in different areas of memory or in different filesystems. Such configuration of high-level replication provides the dislocation of digital resources within a single archive. It is possible to areas of memory or filesystems are carried out in different technologies. Such dislocation provides technological diversification while recording, which further enhances the reliability of storage. High level replication in the archive means that the resource is stored in several copies of the same archival package, located at different physical addresses (e.g. in a variety of filesystems or on different paths in distributed filesystem). Dislocation within the same archive provides more means of security, because beyond the data replication also provides:

- doubling the security measures of the package,
- diversification of the technical representation of the package when it is stored in different filesystem technology,
- spatial dispersion, as the package is replicated in filesystems located at different physical locations.

1.2 Replication in Federated Archives

Dislocation in separate archives is based on the assumption of possible loose coupling between interacting archives. Such an assumption allows interoperability both between separate archives of the same types, as well as between different types of archives.

Detailed assumptions of such cooperation are as follows. Each of the archives has information about which of federated archives have the resources corresponding to the resources in the archive. To recognize two pieces of digital resource on the federated archives as "corresponding", not necessarily mean identity of the resources. Moreover, the files in the resources also does not have to be bitstream identical. To recognize the digital resource as "corresponding" means the suitability of the information, allowing to replace one digital resource with the other. The archives have to periodically exchange with the information on the state of resources, and on their damages. Note, that replacing the digital resource with the corresponding one can cause some lack of information. The degree of resources correspondence should be determined in formalized, qualitative manner. Archives must have information on how to interact to identify the corresponding resources. Archives should regularly exchange information on the state of the resources, in particular, report a problem with one.

It should be emphasized that the above-described adequacy of resources does not need to concern the whole package, but that resources, i.e. those parts of the package, which are logically separated, entirely independent and have their own separate metadata.

So defined dislocation requires ensuring communication with other interacting or federated archives, which our archive knows that store resources corresponding to our own resources. The following problems need to be solved:

- how to identify local resource counterparts in other archives;
- how to coordinate action on dislocated resource to increase global storage reliability;
- how to ensure communication between different types of archives.

2   Communication Among Interacting Archives

Since the dislocation in many archives may take place in the federation system
with different structure and different principles of action, links between interacting
archives must be as small as possible. Moreover, the communication ought to base on
commonly used technology and do not impose significant additional requirements for
cooperative systems. It would be desirable even to activities related to this interaction
may be needed in some archives done manually, or at most semiautomatically.

2.1   Acquisition and Placement Information
About the Corresponding Resources

Due to the above described restriction it is assumed that the exchange of information
of a corresponding resource in the archives will interact, at least initially, be made
by non-technical means. This information in the archive can be associated with data
as a specific type of metadata.

You can create mechanisms to identify the federated archives resources that may
correspond to one, e.g. on the basis of global identifiers such as URN or DOI. This
process can be fully automated, but only in the case of bitstream identical copies of
resource. Regarding the content identity is necessary to assess the compatibility of
resources by a human, and so the process is semi-automatic.

The automated replacement of the resources is not anticipated. Due to the potential
difference in technological archives, legal and organizational problems, the automation of such a process can cause significant difficulties. If necessary, a resource repro-
duction among a number of archives may be carried out in conventional manner by
performing the Outgest operation in one of archives and introducing the resulting
copy of the resource to the other archives using their own Ingest operations.

2.2   Identification of the Archives

Identification of the individual archives in the archives federated network will be
based on the URI addresses. Corresponding resources will be federated archives
identified by placing in each of the archives contains a list of resources:

- internal resource identifier,
- identifier of interacting archive and resource identifier in that archive, in accordance
  with the identification system shared by that archive,
- qualitative measure of adequacy, indicating whether the corresponding resource
  is an exact copy, or has a lower affinity with a local resource.

If cooperating archive provides for their resources different ways to identify, it is proposed to select one of the commonly used global unique identifiers, e.g. URN, DOI. For communication between archives we propose to use the resource identifiers specific to the called archive, made available by that archive. These identifiers will be stored in URI format, with potential application of the scheme Info URI scheme.

2.3 Archives Communication

Communication between interacting archives should have two functions:

- periodic exchange of information confirming the existence and good condition of corresponding resources,
- coordination of cooperating archives in case of the need for making resource operations involving the risk.

This communication should be based on simple and universally understandable ways of information exchange. XML features cause that it seems to be the ideal means to this type of communication. It is assumed, therefore, that the communication between the archives will be based on the exchange of XML messages.

The way in which messages will be generated and transmitted between archives need not to be uniform and may differ significantly in various archives from full integration with the archive, automation and messaging e.g. using Web services, by supporting communication for additional software featuring the archive, to the manual operation, where the software only facilitates the reading of the message and to formulate answers, and the rest of the actions require acting of an archive operator or administrator.

3 Coordination Methods in the Federated Archives

To develop dislocation of a digital resources, a loosely coupling among federated archives should be assumed. An agreement between the federated archives should be as simple as possible: only communication details (message format, means of message passing) and minimal set of exchanged information should be required.

3.1 Dislocation Management

When you want to perform operation on the resource, which poses a risk of damage or loss of data, then you should protect other copies of this resource before performing on them risky operations at the same time. One solution is to perform dangerous operations on a single copy of the resource at a time, but because of the loose coupling between archives, this solution is technically difficult. In connection with this the archive wishing to perform the risky operation should inform other archives of its intention to implement such an operation. Archive, which receives such information should not perform risky operations on data resources within the specified time.

The risky operations on the digital resource are e.g.:

- magnetic renewal of the data carrier that holds the part of the digital resource,
- relocation—physical transfer of the digital resource onto the other data carrier,
- logical migration of the data (i.e. format change of at least one of the files of digital resource).

Archive planning risky operation should, before it begins, to make sure that the counterparts of resources on which to perform this operation are still properly stored in the federated archives. To confirm this it is enough to get the information that the correct storage counterparts was checked in reasonably recent time.

Dislocation Within a Single Archive

This is a special case of high-level replication. Archive should not carry out risky operations on different copies of the same package at the same time.

Dislocation in Federated Archives

In this case we are dealing with a distributed (multi-agent) problem reservation resource. There are algorithms for dedicated operating systems for solving this problem.

3.2 Review of Existing Algorithms on Resource Reservation in Distributed Systems

The archives are spatially dispersed, have partial knowledge, different objectives, take decisions for their own goals. This all makes you treat the archives (and more specifically their systems of dislocation management) as agents. If the resource is stored in different archives, we can treat it as a critical resource (actually quasi-critical, since it is assumed that a resource rather should not be required to access by more than one archive simultaneously). The remainder of the section presents a brief overview of algorithms for resource reservation in distributed systems. We present the algorithms according to the archive terminology, where a processor is an archive.

3.2.1   Lamport Algorithm

Lamport algorithm [4] is the simplest algorithm of resource reservation in distributed systems. It is assumed that for each resource every agent has a queue for reservation requests for a resource. Queues are ordered according to the timestamps.

Requesting archive process:

- enter the request to the queue,
- send the request message to all the archives,
- wait for answers from all the archives,
- if the current request is on the beginning of the queue, and all archives agreed to the request, reserve resource,
- after the resource is released, delete request from the queue and send the release message to all the archives.

Remaining archives:

- after receiving the request message, enter it to the queue, and answer with the timestamp,
- after receiving the release message, delete corresponding request from the queue.

3.2.2 Ricart-Agrawala Algorithm

Ricart-Agrawala algorithm [9] expands the Lamport algorithm.

Requesting archive sends request to all the archives, the request contains identifier of the requesting archive, and the timestamp.

After receiving the request, the archive immediately answers with timestamp, iff (i) archive is not interested in resource, or (ii) archive has a newer timestamp. Otherwise, the archive refuses the request, and requesting archive cannot get the access to the resource.

The requesting archive takes the operations on the resource iff it receives positive answer from all remaining archives. After the archive takes operations on the resource, it sends message on release the resource to all the archives.

The algorithm is susceptible to lack of answer from any archive. Where it is assumed that each of archive always reply then the algorithm may be used, however such assumption cannot be met, as the archives are loosely coupled.

3.2.3 Meakawa Algorithm

The Meakawa algorithm [5] assumes, that to get access to the resource, the archive does not need answer from all the remaining ones. Archive i does not need permission from all the archives that have the resource, but from their subset R_i. The operation of this algorithm can be summarized as follows: archive can allow you to lock a resource only when it has not granted it before to another archive.

- Archive who requested blocking the resource block, sends the request $Request(i, resource)$ message to all the archives of which requires a permit—subset R_i.
- When the j archive receives message $Request(i, resource)$, it sends answer $Inform(resource)$ to archive i, only when it has not send such message since

receiving the last message *Release(resource)*. Otherwise, the message *Request(i, resource)* goes to the queue.

- When archive *i* receives answer from all archives from subset R_i, it can block the resource and perform the risky operation on it.
- When archive *i* finishes performing the risky operation on the resource, it sends *Release(resource)* message to all archives from subset R_i. When archive *j* receives the message, it sends *Inform(resource)* message to following archive, that waits in the queue (its message *Request(k, resource)* waits in queue) and removes the message from the queue. If the queue is empty, the archive updates its state.

There are four certain conditions for the creation of subsets R_i:

M1 $\forall_i : \forall_j R_i \cap R_j \neq \emptyset$—subsets for various archives overlap,
M2 $\forall_i : i \in R_i$—each archive belongs to a subset,
M3 $\forall_i : |R_i| = K$—the size of each subset R_i is K,
M4   any archive *j* is included in K subsets.

The M1 and M2 conditions are required to proper operation of the algorithm. The M3 and M4 conditions are not required, make equal division of work in the algorithm.

High level dislocation applying the Meakawa algorithm requires additional parameters:

- remember your subset of the resource block (for each resource),
- for each package, the archive must set the requests queue,
- remember, if the resource archive allow the sending of permits blocking.

In addition, you must specify an algorithm to create subsets locking resources. The possibilities are as follows.

- The subsets R_i^p are treated as the sets of archives, that holds the resource *p*; when resource *p* is ingested into archive *k*, it broadcasts a message to all the remaining archives. The archives that also store the resource *p*, add the archive *k* to their subset, and answer to archive *k* the information on storing resource *p*. Archive *k* adds all the archives, that answered to its subset.
- The subsets R_i^p are the subsets of archives, that holds the resource *p*; when the resource *p* is ingested into archive *k*, the procedure is the same as above. After setting all the owners of resource *p*, archives delete some elements from their subsets, and check whether the constraints Mx hold.

3.2.4   Other Algorithms

Let us briefly analyze other algorithms. The Suzuki-Kasami algorithm [10], assumes the existence of the token, which holder may reserve a resource. On the other hand, the Raymond algorithm [8] assumes a hierarchy of agents. Both of these requirements can not be achieved in an archives federation.

4 Operation Coordination on Dislocated Resources

4.1 Risky Operation Coordination Through Interaction Protocol

As the implementation of the protocol coordinated risky operations, we propose to modify the Ricart-Agrawala algorithm [9]. To adjust the algorithm to the needs of coordination of digital data, consider the lack of response to requests. We assume, that when an archive does not answer to the request, it is interpreted that it is not interested in the resource. It is therefore proposed that, after a period of time determined to get an answer from another archive, the lack of response is treated as a positive answer.

The coordination protocol for resource reservation is depicted in Fig. 1, using AUML (Agent UML [2]) interaction diagram. An interaction protocol acts as follows.

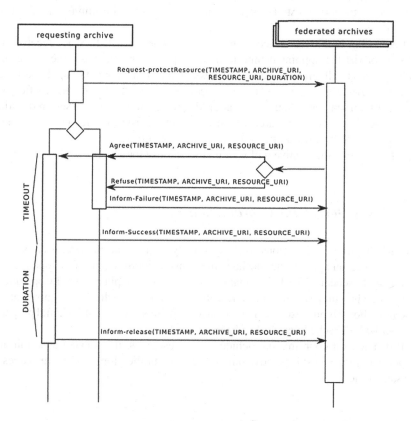

Fig. 1 AUML interaction diagram for coordination protocol of resource reservation

Archive requesting for a resource sends request `Request-protectResource` (`TIMESTAMP, ARCHIVE_URI, RESOURCE_URI, DURATION`) to all the archives, where:

- `TIMESTAMP`—timestamp,
- `ARCHIVE_URI`—requesting archive global identifier,
- `RESOURCE_URI`—local identifier of a requesting resource,
- `DURATION`—the time for which the archive wants to reserve a resource.

After receiving the request, the archive answers immediately positive with the timestamp (`Agree(TIMESTAMP, ARCHIVE_URI, RESOURCE_URI)`) iff: (i) archive is not interested with resource, or (ii) archive has lower priority (timestamp). Otherwise, when archive is interested with resource, and has a higher priority, it refuses the request (`Refuse(TIMESTAMP, ARCHIVE_URI, RESOURCE_URI)`.

When requesting archive receives at least one refuse, it does not get access to resource, and have to send to all remaining archives the message `Inform-Failure(TIMESTAMP, ARCHIVE_URI, RESOURCE_URI)`, informing of failure of the request.

Requesting archive gets access to resource, when: (i) it does not receive any refuses, or (ii) the appointed time (`TIMEOUT`) passes. In this case, the requesting archive sends to all remaining archives message informing about taking the resource over: `Inform-Success(TIMESTAMP, ARCHIVE_URI, RESOURCE_URI)`. After taking operation on the resource, or after the reservation time passes, requesting archive sends to remaining archives message on releasing the resource: `Inform-release(TIMESTAMP, ARCHIVE_URI, RESOURCE_URI)`.

It follows that the archive will wait up to `TIMEOUT` for a response.

4.2 Risky Operations Coordination

As CREDO archive operations working directly on archival filesystems are subject to scheduling, an appropriate mechanism to ensure coordination of the operation is to transfer to scheduler such information, which will force planning risky operations at a time, when they will not interfere with operations on dislocation equivalents of resources. For this purpose, Archive Management Subsystem (AMS) should properly interact with the scheduler (part of PMS).

If demand for coordination includes many resources, it is treated as a unit and response to the request is positive only if it can be fulfilled for all of the resources at the same time.

4.2.1 Coordination of Own Operations

When the archive is planning the risky operation on the resources, the archive should work as follows.

- Undergo risky operation scheduling.
- However, do not start the operation, but before the start synchronize the resources with federated archives.

  - Confirm that in the federated archives counterparts of processed resources have the correct state verified.
  - Send federated archives information about the planned risky operations on the resources, along with the expected start time and expected operation duration.

- If there are no conflicts, run operation at the scheduled time.
- If conflicts occur, set the earliest start time of the operation, so that it began after the scheduled end of the operation in other archives and request for rescheduling.

4.2.2   Coordination of Foreign Operations

In the event that the archive receives notice from another archive of the need to perform a risky operation on resources, the archive should work as follows.

- Check whether in time declared by those archives, risky operations on the resource are not planned in our archive.
- If they are, perform the interaction protocol for risky operation coordination (see Sect. 4.1).
- If not, send an approval for confirmation.
- Set the estimated duration, and scheduled start time of the operation.

This procedure will ensure that the scheduler, in time at which the federated archive plans to carry out risky operations, does not schedule the risky operation on own resource.

Since the above mentioned security stems from a very far-reaching precautions, and there are many other mechanisms to ensure the correctness of the data, it is not necessary to check whether the actual execution times of operation are still valid.

5   Implementation Proposal

As in the archive federation particular archives act autonomously, have its own goals and preferences, we propose to look at them as the agents, and at the federation of archives as the multi-agent system. Specifically, the module responsible for communication with the other archives, can be modeled as the agent. Furthermore, the implementation of the multi-agent environment to communicate the federated archives is needed.

Agents communicate using coordination protocol of resource reservation (see Fig. 1). Agents are FIPA-compatible, where FIPA (The Foundation for Intelligent Physical Agents) is an organization that promotes agent-based technology and the interoperability of its standards with other technologies [6]. To implement the agents on each of the archives, we propose to use JADE (Java Agent DEvelopment Framework), a FIPA-compliant agent framework [1] for implementation of multi-agent system, as the JADE is implementation of the FIPA standards. To avoid problems with running the JADE on the JVM (Java Virtual Machine) (checking state of the service, re-running, access to network), we propose to run it on the application server. We can run JADE on OSGi architecture provided on the Glassfish application server (from version 3.0). The OSGi technology is a set of specifications that define a dynamic component system for Java. These specifications enable a development model where applications are (dynamically) composed of many different (reusable) components [7]. The OSGi allows, among others, to remotely install, start, stop, update, uninstall the service without rebooting. To run JADE using the OSGi, we can use JADE-OSGi library which allows for it. The next issue is the communication of agents. Normally, the agents communicate through a TCP/IP port (1099 by default), however this may post a threat of attacks. The other solution is to connect the federated archives using VPN. The another solution is to communicate the agents using Web Services. The technology to integrate the agents with the Web Services is WSIG (JADE Web Service Integration Gateway) [3]. It allows to invoke agents from Web Service clients.

6 Conclusion

The dislocation of digital resources is one of high-level methods to ensure reliable digital data preservation. Moreover, it creates possibility for semi-automatic reconstruction of damaged resources. To apply dislocation in multi-archive environment, the federation of—possibly heterogeneous—archives, that inter-operate with each other, is needed.

References

1. Bellifemine, F., Poggi, A., Rimassa, G. Developing multi-agent systems with a FIPA-compliant agent framework *Software-Practice and Experience* 31(2):103–128, 2001.
2. Bauer, B., Müller, J.P., Odell, J. Agent UML: A formalism for specifying multiagent software systems *International journal of software engineering and knowledge engineering*, 11(03):207–230, World Scientific, 2001.
3. Jade Board. Jade web services integration gateway (WSIG) guide *Telecom Italia*, 2008.
4. Lamport, L. Time, clocks, and the ordering of events in a distributed system. *Communications of the ACM*, 21(7):558–565, 1978.
5. Maekawa, M., Oldehoeft, A.E., Oldehoeft, R.R. *Operating Systems: Advanced Concept*. Benjamin/Cummings Publishing Company, Inc., 1987.
6. O'Brien, P.D., Nicol, R.C. FIPA – towards a standard for software agents *BT Technology Journal*, 16(3):51–59, 1998.
7. Open Services Gateway initiative http://www.osgi.org/developer/architecture.

8. Raymond, K. A tree-based algorithm for distributed mutual exclusion. *ACM Transactions on Computer Systems (TOCS)*, 7(1):61–77, 1989.
9. Ricart, G., Agrawala, A.K. An optimal algorithm for mutual exclusion in computer networks. *Communications of the ACM*, 24(1):9–17, 1981.
10. Suzuki, I., Kasami, T. A distributed mutual exclusion algorithm. *ACM Transactions on Computer Systems (TOCS)*, 3(4):344–349, 1985.

Index

© Springer International Publishing AG 2017
T. Traczyk et al. (eds.), *Digital Preservation: Putting It to Work*,
Studies in Computational Intelligence 700, DOI 10.1007/978-3-319-51801-5

Printed in the United States
By Bookmasters